SpringerBriefs in History of Science and Technology

Series Editors

Gerard Alberts, University of Amsterdam, Amsterdam, The Netherlands

Theodore Arabatzis, University of Athens, Athens, Greece

Bretislav Friedrich, Fritz Haber Institut der Max Planck Gesellschaft, Berlin, Germany

Ulf Hashagen, Deutsches Museum, Munich, Germany

Dieter Hoffmann, Max-Planck-Institute for the History of Science, Berlin, Germany

Simon Mitton, University of Cambridge, Cambridge, UK

David Pantalony, Ingenium - Canada's Museums of Science and Innovation / University of Ottawa, Ottawa, ON, Canada

Matteo Valleriani, Max-Planck-Institute for the History of Science, Berlin, Germany

The *SpringerBriefs in the History of Science and Technology* series addresses, in the broadest sense, the history of man's empirical and theoretical understanding of Nature and Technology, and the processes and people involved in acquiring this understanding. The series provides a forum for shorter works that escape the traditional book model. SpringerBriefs are typically between 50 and 125 pages in length (max. ca. 50.000 words); between the limit of a journal review article and a conventional book.

Authored by science and technology historians and scientists across physics, chemistry, biology, medicine, mathematics, astronomy, technology and related disciplines, the volumes will comprise:

1. Accounts of the development of scientific ideas at any pertinent stage in history: from the earliest observations of Babylonian Astronomers, through the abstract and practical advances of Classical Antiquity, the scientific revolution of the Age of Reason, to the fast-moving progress seen in modern R&D;
2. Biographies, full or partial, of key thinkers and science and technology pioneers;
3. Historical documents such as letters, manuscripts, or reports, together with annotation and analysis;
4. Works addressing social aspects of science and technology history (the role of institutes and societies, the interaction of science and politics, historical and political epistemology);
5. Works in the emerging field of computational history.

The series is aimed at a wide audience of academic scientists and historians, but many of the volumes will also appeal to general readers interested in the evolution of scientific ideas, in the relation between science and technology, and in the role technology shaped our world.

All proposals will be considered.

Hannah C. Erlwein · Katja Krause

Revisiting Premodern Islamic Science and Experience

MAX PLANCK GESELLSCHAFT

Springer

Hannah C. Erlwein
Max Planck Institute for the History of Science
Berlin, Germany

Katja Krause
Max Planck Institute for the History of Science
Berlin, Germany

Technische Universität Berlin
Berlin, Germany

ISSN 2211-4564 ISSN 2211-4572 (electronic)
SpringerBriefs in History of Science and Technology
ISBN 978-3-031-76084-6 ISBN 978-3-031-76085-3 (eBook)
https://doi.org/10.1007/978-3-031-76085-3

Open Access funding is provided by the Max Planck Society and the Max Planck Institute for Comparative and International Private Law

This Springer imprint is published by the registered company Springer Nature Switzerland AG
The registered company address is: Gewerbestrasse 11, 6330 Cham, Switzerland

The original version of the book has been revised. A correction to this book can be found at https://doi.org/10.1007/978-3-031-76085-3_6

Acknowledgments

This volume has its origins in a workshop convened at the Max Planck Institute for the History of Science (MPIWG), Berlin, in March 2022. "Premodern Islamic Science: Demarcating Experiential Knowledge and Scientific Knowledge" was a forum that examined the intersections of Islamic science and experience across the disciplines of philosophy, theology, medicine, and astronomy/astrology. We thank all the colleagues who participated, including those who were unable to contribute to the final volume. In particular, we acknowledge the contributions of Livnat Holtzman, Miriam Ovadia, Elvira Wakelnig, Miriam Shefer Mossensohn, and Robert Morrison. Their presentations were not only thought-provoking, but have indelibly shaped the conception and content of this work.

We also extend our deep appreciation to Steven Harvey, whose insightful discussions about the project in its entirety, coupled with invaluable comments on the Prologue, have enriched it immeasurably.

Profound thanks are due to the MPIWG. We are especially grateful for the assistance of the Library and of Kate Sturge, the publications manager in Katja Krause's research group "Experience in the Premodern Sciences of Soul and Body." Kate's dedication and meticulous editorial work were critical in ensuring the clarity and elegance of this book. The Open Access publication of *Revisiting Premodern Islamic Science and Experience* was kindly funded by a grant from the MPIWG Open Access Monograph Publishing Fund.

Lastly, we wish to convey our deepest gratitude to those closest to us, whose unwavering support has been the foundation of our endeavors. Hannah thanks her parents and her husband, Tom, for their enduring encouragement. Katja honors the memory of her mother, who passed away during the making of this volume, and expresses heartfelt gratitude to her family, Frank and Elisabeth, for their steadfast love, trust, and support. Their presence has been a wellspring of inspiration and strength throughout this process.

Contents

Chapter 1
Prologue: Elements of a New Architecture of Experience and Science

Hannah C. Erlwein and Katja Krause

Abstract *Revisiting Premodern Islamic Science and Experience* seeks to answer two questions: *What kind of* experience constituted premodern Islamic science? And *in what ways* did that experience constitute science? Our answer to these questions is not reducible to experience as empirical method or practice; neither the empirical method nor any empirical practice exhaust the scope of premodern experience in its relevance for science. To illuminate that argument, we turn our gaze to where experience naturally arises: within the historical scientist. Specifically, we address what we call the subject-rootedness or subject-dependence of the scientist's experience, be that scientist-subject historically real or idealized. By founding our story on the scientist—and the ways in which his experience was rooted in acquired skill, habit, expertise, and practice—we offer the history of science a solid foundation on which to build its analyses of premodern science and the scope and role of experience therein. Bracketed by an introduction that advances a new theoretical framework and an epilogue that reflects on the findings and their implications, this volume presents its argument through three chapters, each of which investigates the place of experience in one branch of premodern Islamic science: astrology, philosophical psychology, and theology.

Keywords Experience · Islamic science · Empirical method and practice · Internalized objectivity · Epistemic community · Subject-rootedness and subject-dependence · *Tajriba*

H. C. Erlwein (✉) · K. Krause
Technische Universität Berlin, Berlin, Germany
e-mail: hannah.erlwein@gmail.com

K. Krause
e-mail: kkrause@mpiwg-berlin.mpg.de

K. Krause
Technische Universität Berlin, Berlin, Germany

H. C. Erlwein and K. Krause, *Revisiting Premodern Islamic Science and Experience*, SpringerBriefs in History of Science and Technology,
https://doi.org/10.1007/978-3-031-76085-3_1

1.1 Laying the Foundations

This is a book about experience in premodern Islamic science. It seeks to answer two questions: *What kind of* experience constituted premodern Islamic science? And *in what ways* did that experience constitute science? Our answer to these questions will not be reducible to experience as empirical method or practice, because we contend that neither the empirical method nor any empirical practice exhaust the scope of premodern experience in its significance for science. To illuminate that argument, we will undertake a shift of perspective and turn our gaze to the place where experience naturally arises: within the historical scientist. We highlight the internalizations of experience that the scientist—be he a historically real or idealized subject—comes to possess as he acquires and uses experience in his scientific endeavors. More broadly, we argue that this approach to the scientist, his internalized experience, and the resulting transformations of his scientific practice can supply the history of science with a solid foundation on which to build its analyses of premodern science and the scope and role of experience therein.

Premodern Islamic science presents a perfect framework for this larger project, for three reasons. First, as we discuss in this Prologue, because existing historiography requires a new architecture of experience and of science that is grounded in the subjects who embodied and ensouled them. Second, because the experiences and scientific outputs of premodern Islamic scientists have left a lasting legacy in many spaces near and far (though this legacy is too large to be investigated here). And third, because Islamic scientists have left behind particularly abundant historical traces of different kinds of experience and science for us to investigate. These traces are what our book presents in its three main chapters and an epilogue that reflects on the findings by adding its own considerations and evidence.

It is not our aim here to argue for some kind of "Islamic exceptionalism," contrasting the ways in which Islamic scientists thought about scientific experience with some sort of "Western" way of scientific thinking. Rather, we wish to show that there is a divide between premodern and modern ways of thinking about science in general and scientific experience in particular. As a result, the insights presented in this book may possess the potential to transcend the specific context of Islamic science and contribute to a much broader discourse on the nature of experience, science, and society.

1.2 Surveying the Historiography of Premodern Science

The new intellectual architecture we advocate, with its emphasis on the scientist-subject, gains significance when viewed against the backdrop of recent historiography of premodern science. A brief survey of this historiography exposes the frameworks and categories that historians have employed to study premodern science and experience, while also indicating their limitations when seeking to understand experience in premodern Islamic science.

One of the most widespread assumptions in the histories of science that appeared up to approximately the 1970s was that, in some significant way or another, experience gives rise to scientific knowledge. This view was predominantly based on framing the relationship of science and experience as a relationship of method: experience is the method that brings about scientific knowledge. In turn, that assumption relied on a largely *internalist* perspective, asking how it is that experience is able to yield scientific knowledge and studying how the relationship between the two developed throughout history.[1] For some of these classic histories of science, experience and science are so inextricably linked as to justify the claim that modern science *in toto* was born when, in the wake of the Scientific Revolution, the empirical method rose to the unparalleled position of historical and scientific prominence that it holds to this day.[2]

More recently, social and practical turns in the history of science cast doubt upon the exclusively internalist approach to premodern science, and instead *externalist* histories began to emphasize historical contexts. As a consequence, the internalist assumption that experience as an empirical method gives rise to science became unviable, and the many ways in which historical circumstances shaped the relationship between premodern science and experience came to be taken into account.[3] Somewhat paradoxically, however, this shift in perspective from method to (social) practice did not affect the familiar narrative that experience gives rise to science.[4]

---

[1] Classic applications of this assumption are Duhem, *To Save the Phenomena*; Thorndike, "Roger Bacon and Experimental Method," which laid the groundwork for his later eight-volume *magnum opus*, Thorndike, *History of Magic and Experimental Science*; Dijksterhuis, *Mechanization of the World Picture*; Crombie, *Robert Grosseteste*; Maier, *Die Vorläufer Galileis im 14. Jahrhundert*; Maier, *Zwischen Philosophie und Mechanik*; Clagett, *Science of Mechanics in the Middle Ages*; Grant, *Foundations of Modern Science*. Outside the history of science, too, this was a paradigm, as is evidenced by, for instance, Ernst Mach's 1896 book *The Analysis of Sensations*. This type of history is continued in philosophical studies on premodern experience to this day; see, for example, Hossfeld, *Albertus Magnus als Naturphilosoph und Naturwissenschaftler*; Jacquart, "Die Medizin als Wissenschaftsdisziplin und ihre Themen"; Dear, "Meanings of Experience."

[2] Thus Butterfield, *Origins of Modern Science*, vii–viii: "Since that revolution overturned the authority in science not only of the middle ages but also of the ancient world [...], it outshines everything since the rise of Christianity and reduces the Renaissance and Reformation to the rank of mere episodes [...]. [It] looms so large as the real origin both of the modern world and of the modern mentality that our customary periodisation of European history has become an anachronism and an encumbrance." Similar views were propagated by Duhem, *L'évolution de la mécanique*, and Crombie, *Robert Grosseteste*.

[3] Some of the most important second-order considerations on the internalist/externalist divide in the history of science are Basalla, *Rise of Modern Science*; Lakatos, "History of Science"; Bynum, Browne, and Porter, *Dictionary of the History of Science*, s.vv. "Externalism," "Internalism"; Shapin, "Discipline and Bounding."

[4] The most famous example is Shapin and Schaffer, *Leviathan and the Air-Pump*, which investigates "the historical circumstances [...] in which experimental practices became institutionalized" (3) and "method understood as real practical activity" (14). See also, in some parts, Gal, *Origins of Modern*

The epistemic values that the internalist histories had associated with experience were thus simply translated into the externalist histories without being revised in substantial ways: empirical method became empirical practice.

The new historiographies shifted their attention away from theoretical constructs of empiricism and worked hard to uncover the practices of empiricism in many different spaces, but especially as they were tied to the two senses of sight and touch.[5] Such externalist histories, which have overwhelmingly focused on early modern Western Europe, concluded that empirical practice (in their definition) did not belong to the old, premodern *scientia* of the learned elites, who are frequently described as "bookish" in the literature, but to more popular and widespread artisanal endeavors.[6]

Partly as a consequence of the growing sentiment among historians of science that our discipline should be steered into supposedly more democratic waters and turn away from narrowly focusing on the *scientia* of the elites, the history of science expanded into the history of knowledge. With this last "turn," the relevance of the discipline's most vexed questions about what counts as science throughout history and what distinguishes scientific knowledge from other forms of knowledge, such as everyday knowledge, has faded.[7] The question of how science relates to experience thus takes a back seat—not because experience has been written out of our discipline, but because science has.[8]

Science. However, Gal already goes beyond these historiographies, acknowledging, for instance, that magic was a tradition that at times competed, at times collaborated with premodern science.

[5] Examples include Shank and Lindberg, "Introduction"; Eamon, *Science and the Secrets of Nature*; Pomata and Siraisi, *Historia*; Kusukawa and Maclean, *Transmitting Knowledge*; Young, "Experimentalist as Spectator"; Daston and Park, *Wonders and the Order of Nature*; Dear, "Meanings of Experience"; Leong, *Recipes and Everyday Knowledge*; Rankin, *Poison Trials*; Werrett, *Thrifty Science*; Truitt, *Medieval Robots*.

[6] A typical example of the treatment of premodern scientists in light of experimentalist historiographies is Barry Allen's evaluation of Grosseteste's scientific approach: "He was 'too bookish' for experiments [...] and never contrived an experiment for the explicit purpose of testing a hypothesis." Allen makes these claims based on McEvoy's *Philosophy of Robert Grosseteste* and Dales's "Grosseteste's Scientific Works" rather than the primary source. Allen, *Empiricisms*, 458n25. Similar disparaging judgments in the literature are also made about Albert the Great. To name just one recent example, in the standard reference work for medieval philosophy in German, Peter Schulthess writes: "Albert the Great argues that confirmation through experience, which is indispensable in natural philosophy, offers greater certainty and is more beneficial than mere intellectual consideration and deduction. Here, 'experimentum' must be understood, as Aristotle does, more as experience in general than as a planned experiment in a controlled setup; it often involves an eclectic selection of examples from one's own experience and from experiences of others drawn from literature. [...] Since this form of empiricism does not involve direct, real, regular, and planned observations, it can be called 'empiricism without observation.'" Schulthess, "Methode in der Naturphilosophie," 1363–1364. Here and throughout, all translations are our own unless otherwise attributed.

[7] Among the contributions that catalyzed this turn in the history of science are Daston, "History of Science"; Renn, "History of Science"; Dupré and Somsen, "History of Knowledge."

[8] The Max Planck Research Group "Experience in the Premodern Sciences of Soul and Body," led by Katja Krause, aims to return to solid, properly historicized categories of experience and science. See Krause with Auxent and Weil, "Introduction: Making Sense of Nature in the Premodern World," which anticipates the present volume's discussion of science and experience in the premodern world

1.3 Surveying the Historiography of Premodern Islamic Science

Parallel to these trends in historiographies of premodern science—which often, though not always, focused on European developments and yet displayed certain universalizing ambitions—historians of the science and philosophy of the Islamic world have expressed a similar general view: in premodern Islamic science, they argued, it was experience in the empirical sense that gave rise to scientific knowledge.

A plenitude of studies has shown that the disciplines canonically studied in the history of science—medicine, astronomy, and optics—relied heavily on experience in both their method and their practice. Studies in the history of philosophy, moreover, described explicit discussions about experience by Islamic scientists.[9] Some historians were even eager to show that premodern Islamic scientists *already* employed the empirical method and as such were no less empirical than their early modern counterparts in Europe. Historical actors such as Ibn Sīnā (d. 427/1037) were discovered to be empiricists through and through, and said to have formulated not just any empiricism, "but an empiricism [...] in the Western European sense as it was later to appear in the work of John Locke."[10] Other historians found the very origins of the empirical method itself in premodern Islamic science, and declared: "It is a mistake to suppose that the experimental method is a European discovery. [...] Europe has been rather slow to recognize the Islamic origin of her scientific method."[11] Others again took a more skeptical view on the significance of the empirical method in premodern Islamic science:

> The empiricism which came to be regarded in the West as one expression of scientific enquiry was completely lost on the Muslims. Western science realized the value of empiricism and succeeded in sifting it out of theology and metaphysics. [...] The Muslims, on the other hand, were not able to see [...] the extraordinary potential of [the] empiricist methodology.[12]

Yet even among those critical voices, the empirical method and practice were made the standard against which Islamic actors were measured. This clearly reflects the trajectory of the historiography of early modern European science that highlight the

---

by conceiving of experience as going beyond the empirical in scope and as principally rooted in the scientist as a human subject.

[9] Consider especially the many articles on experience (especially *tajriba*) in Ibn Sīnā's scientific thought, e.g., Anawati, "Ishām Ibn Sīnā fī taqaddum al-ʿulūm"; Gutas, "Empiricism of Avicenna"; Gutas, "Certainty, Doubt, and Error"; Janssens, "'Experience' (*tajriba*)"; Karimullah, "Avicenna and Galen"; McGinnis, "Scientific Methodologies in Medieval Islam"; McGinnis, "Avicenna's Naturalized Epistemology." See also Klein-Franke, *Vorlesungen*; Rosenthal, *Science and Medicine*; Saliba, "Theory and Observation in Islamic Astronomy."

[10] Gutas, "Empiricism of Avicenna," 393–394.

[11] Iqbal, *Reconstruction of Religious Thought*, 103. Iqbal goes on to quote Robert Briffault, *The Making of Humanity* (201), to support his argument: "Discussions as to who was the originator of the experimental method [...] are part of the colossal misrepresentation of the origins of European civilization."

[12] Hallaq, "Introduction," xlix.

importance for the rise of modern science of the empirical method and empirical practice in the study of nature. Those histories suffer from two key problems, which stem from their emphasis on experience as method or practice.

The first problem is that they unduly limit the *object* of experience to one that is always causally dependent upon direct and external sense perceptions, with or without the mediation of instruments. Only what is externally perceived can be accepted into the historical category of "scientific experience" or "scientifically relevant experience." In contrast, what is *internally* perceived—a category of significance for our historical actors and ubiquitous in their considerations on the soul—makes almost no appearance in these historiographies at all.

The second problem is that they consistently seek those epistemic *functions* of experience that resemble observations and experiments as they were conceived in early modern Europe. George Saliba, for instance, stressed that Islamic astronomers used "fresh observations" to critique aspects of Greek astronomical knowledge or the very "methods of observation" they had inherited from the Greeks.[13] And Felix Klein-Franke noted of Islamic medicine: "In addition to dogmatic medicine, which reached specific conclusions from theories in a deductive manner, there was also the converse process, reaching general findings from sensually perceptible phenomena in an inductive manner."[14]

This emphasis on experience as empirical method or practice constructs experience as something to be viewed through the lens of function alone: how much, how often, how direct, how good, how precise, how trustworthy, and so on.

More than a hundred years ago, however, Ernst Cassirer eloquently made his case in *Substance and Function* that premodern concepts of science fundamentally differ from modern ones—and we ask, as a consequence, how modern historiography can properly account for this. As Cassirer showed, premodern science, in most of its expressions, revolved around the concept of substance, whereas modern science is characterized by the precedence it gives to the concept of function:

> The [Aristotelian scientific] process of comparing things and of grouping them together according to similar properties, as it is expressed first of all in language, does not lead to what is indefinite, but if rightly conducted, ends in the discovery of the real essences of things. Thought [in Aristotelian science] only isolates the specific type; this latter is contained as an active factor in the individual concrete reality and gives the general pattern to the manifold special forms.

In contrast, says Cassirer, the modern concept of function—function here understood in its mathematical sense—refers to the various logical relations that are conceivable for the objects of scientific investigation. This approach gives "a universal *rule* for the connection of the particulars themselves" and occasions a very distinct scientific perspective:

> Thus we can conceive members of series ordered according to [the rule of] equality or inequality, number and magnitude, spatial and temporal relations, or causal dependence. [...] We do not isolate any abstract part whatever from the manifold before us [as thinking

[13] Saliba, *Islamic Science*, 132.

[14] Klein-Franke, *Vorlesungen*, 95.

> in terms of substance does], but we create for its members a definite relation by thinking of them as bound together by an inclusive law.[15]

Cassirer's general identification of a distinct premodern scientific goal—one of substance rather than function—can also be applied in investigating the history of Islamic science, even though he was not studying this at the time when he was writing *Substance and Function.* Indeed, we present Cassirer's sharp observation at such length in order to underscore our point that most of the existing historiography implicitly reduces premodern Islamic experience to a problem of empirical method. This reduction makes use of the same kind of *function*-thinking as modern science does, subsuming historical facts under function without critical self-reflection—but it disallows the historical objects' expression of their own ontologies, relations, and workings.

In the case of premodern Islamic science, because many of its fields borrowed heavily from Greek science, which Cassirer had studied intensely as the object of his historiographical argument, it cancels out the all-important ontological grounding of function in substance, and by extension the all-important epistemic resolution of function to substance. Yet this way of thinking needs to inform our historical investigations of premodern experience, too, because it situates function where it naturally belonged for these historical actors: in substance, and not in a method without metaphysics. Inversely, we argue, to analyze premodern scientific experience in the Islamic world primarily in relation to function (in Cassirer's sense) rather than in relation to substance would be a historiographical anachronism.

The implications of this functional anachronism loom large. On the one hand, when experience is conceptualized in this way, its historical trace becomes procedural, operational, and methodological *per definitionem.* On the other hand, and more importantly, a focus on epistemic functions can discern only a limited set of purposes that were served by experience in premodern Islamic science—namely, so we are told in many historical accounts, to attain the first principles of science as the basis for demonstration, to verify or falsify previous experiences recorded in the texts of the tradition, and to generate judgments of certainty as the desired goal of demonstrations.[16]

The general functionality approach thus constructs a merely partial image of the kind of experience that constituted premodern Islamic science, and at the same time disregards the substantial nature of experience that the historical actors themselves assigned to it: the rootedness of all these different functions of experience in its subject, the scientist himself. It is this subject that our historiography of premodern Islamic science puts center stage, making it a foundation and a methodological point of origin—just as premodern Islamic science did itself.

---

[15] Cassirer, *Substance and Function*, 7 and 16–20 (original emphasis).

[16] This is the view presented in studies such as Gutas, "Certainty, Doubt, Error," 279; Gutas, "Empiricism of Avicenna," 398–399; Saliba, *Islamic Science*, 132; Dear, "Meanings of Experience," 108–109 For a valuable alternative approach, see Brentjes, *Teaching and Learning*.

1.4 Elements of a New Architecture

The consequences of the functional anachronism also loom large for our second historical object, premodern science, to which our first historical object, experience, was inevitably tied. In existing histories of Islamic sciences, applying the functional ideal of experience to science—an ideal that was secondary to that of substance in the premodern Islamic context, as we have just shown—has all too often resulted in the almost complete exclusion of some scientific disciplines. In the eyes of historians of science, theology and jurisprudence, for example, have been deemed too "unscientific" to belong to the true canon of the history of science, largely because they did not follow the empirical method or empirical practices resembling those of modern science.[17] Yet for the historical actors it was self-evident that the disciplines they called *ʿilm* (pl. *ʿulūm*), which we here translate as *science*, belonged to the category of the most prized knowledge there was.

Islamic scientists themselves frequently undertook a fundamental division of *ʿulūm*, according to their origin, into "Arabic sciences" and "foreign sciences." The former included religious sciences such as theology, jurisprudence, and grammar. The latter adopted the Greek canon of the arts and sciences, including logic, medicine, astronomy, physics, metaphysics, and practical philosophy. Unsurprisingly, the criteria for this division were subject to debate over time.[18] But as historians, we must ask: When speaking of Islamic science, can we simply ignore those actors' ideas of what does and does not count as science? And can we exclude their discussions and uses of experience within those sciences as irrelevant because they do not share the epistemic values that have so far shaped the historiography of science as described above? Our answer to these questions is a clear "no." This is why astrology, psychology (as the *rational* investigation of the soul), and theology—all considered to be part of the canon of sciences by our historical actors—are the focus of the chapters in this volume. Instead of examining presumedly empirical sciences such as medicine, astronomy, and optics, which have received much attention in existing studies for the reasons just discussed, we deliberately turn to those sciences that have hitherto been regarded as of little relevance by historians of Islamic science and experience.

Revisiting Premodern Islamic Science and Experience therefore begins where histories of premodern Islamic science rarely have: with the scientist—or, perhaps more accurately, the archetype or ideal of the scientist in any given discipline that was called *ʿilm*. That is because the scientist had the most profound impact on the kind of experience that was relevant for attaining scientific knowledge. This turn to

[17] This is exemplified by the title of a series published by Brill, "Islamic Philosophy, Theology and Science," which suggests that philosophy and theology are different from science rather than applying the overarching actors' category *ʿilm*.

[18] Rosenthal, *Classical Heritage in Islam*, 52–70; Akasoy and Fidora, "Structure and Methods of the Sciences."

the acting subject, we propose, results in nothing less than a new architecture of the historiography of premodern Islamic science and experience.[19]

An example may help to corroborate our claim. The premodern Islamic astrologer and the premodern Islamic theologian alike had a special interest in the heavenly bodies. Both regularly observed the movement of the sun, the moon, and the other planets. But one and the same experienceable phenomenon had an entirely different mental significance for the practitioners of these two sciences. The astrologer saw *future* events revealed; the theologian found the *past* event of God's creation of the world. Importantly, it is not that the astrologer and the theologian made use of the same phenomena they experienced, just for different purposes—rather, the theologian experienced the world around him *as* creation, while the astrologer read the world around him *as* an indication of future events.[20] In short, the scientist as theologian had a different experiential relation to the world than did the scientist as astrologer.

Experience in each scientific discipline was deeply shaped and subsequently deployed as much by the cognitive skills and practices as by the goals and aims that the scientist pursued within a particular discipline. These differed between the theologian and the astrologer, rendering the experience of the theologian different from the experience of the astrologer—despite its attaching to one and the same experienced phenomenon. This is also why our historical scientists acknowledged that laypeople could have knowledge about the world thanks to repeated perceptions and experiences, but argued that because laypeople did not partake in the cognitive skills and practices of the community of scientists or share scientists' trained judgment and cultural wisdom, their empirical knowledge did not belong to the realm of scientific knowledge (a point stressed by Charles Burnett in his chapter in this volume). In the eyes of premodern Islamic scientists, there was thus a distinctive, scientifically mediated interplay between the subject of the scientist and the object of his inquiry.

Importantly, however, that interplay does not make the scientist's experience "subjective"—indeed, it defies our modern dichotomy of subjectivity and objectivity altogether. For our historical actors, experience and science amounted to what we might today conceptualize as "internalized objectivity": an objectivity that lived in the subject once it had been defined by the particular configurations of skills,

---

[19] The heirs of Michel Foucault will probably find our approach wanting, because Foucault's aim was to erase the subject. We think, however, that there is no proper premodern history of science without the subject, for the reasons explained in this chapter. Our argumentation aside, there have also been quite recent attempts to draw renewed attention to the subject in Western historiographies, albeit using a very different methodology than ours. Those attempts—most prominently Shapin, *Never Pure*, and Daston and Sibum's special issue "Scientific Personae and Their Histories"—have turned to concrete individuals, their bodies, or the different social personae of the scientist. It is not our aim to follow their methodological paths, which were inspired by material and social approaches to science. Instead, we wish to build a methodology that does epistemic justice to the particular historical actors we are studying. For these actors, as we have argued, the substance (and in our specific case, the subject-substance) of the scientist was the point of origin of experience and science, and it is this that guides us here.

[20] Compare Norwood Russell Hanson's point that all observation is theory-laden. Hanson, *Patterns of Discovery*.

habit, expertise, practice, and knowledge acquired by the scientist within a particular scientific community. In this sense, the subject-rootedness or subject-dependence of experience and science was not the opposite of objectivity but constituted it—and this "internalized objectivity" was an epistemic virtue, not an epistemic vice.

It is on the basis of this kind of internalized objectivity that the pugnacious theologian Ibn Taymiyya (d. 728/1328) was able to argue: "We maintain that knowledge, whether self-evident or acquired, is a matter of relation [*nisbiyya*] and connection [*iḍāfiyya*]. What may be self-evident for one man may be acquired for another."[21] Ibn Taymiyya acknowledged what we here call the subject-dependence of knowledge: the circumstance that different individuals' or groups' epistemic abilities, skills, and practices interact with the object of their knowledge, yet without rendering one relation to knowledge more objective or subjective than another—for instance, the self-evident nature of a proposition versus the rationally deduced nature of a proposition. Different people arrived at knowledge via different paths, but once they had obtained that knowledge, it simply *was* scientific and, now as knowledge that had come to reside in the knower, objective.

For the present-day mind, the concept of internalized objectivity might appear to be a contradiction in terms, since we are so accustomed to seeing a chasm between objectivity and subjectivity—separating subjective arbitrariness from objective universality and the personal, often emotional, interpretations of the world from verifiable facts. Lorraine Daston and Peter Galison have eloquently described that dichotomy, the bread and butter of present-day thought. But they highlight its own historical specificity:

> The mid-nineteenth-century appropriation of the Kantian terminology of objective and subjective in science tended to fuse the epistemological and ethical: the acquisition of knowledge was seen—and felt—to involve a battle of the will against itself. [...] Objectivity was a different, and distinctly epistemological, goal—in contrast to the metaphysical aim of truth. And subjectivity was not merely a synonym for being prone to errors; it was an essential aspect of the human condition, including the pursuit of knowledge. But for mid-nineteenth-century scientists, this epistemological predicament was hopelessly entangled with an ethical one that was also cast in terms of the objective and subjective. To know objectively was to suppress subjectivity, described as a post-Kantian combat of the will with itself.[22]

This duality of subjectivity and objectivity has been omnipresent since the reception of Kant among scientists. But no element of that dilemma haunted either the real or the ideal scientist of the Islamic Golden Age, who did not juxtapose an "untrustworthy scientific self," hopelessly entangled in subjectivity and arbitrariness, with the objective rules of science, which required "self-restraint, self-discipline, self-control."[23] Instead, the premodern Islamic scientist acquired the body of science from the scientific community in which he lived, and subsequently employed that body of science in experiencing the world around him. The community which taught him science ensured that both his science and his experience were far from being

[21] Hallaq, *Ibn Taymiyya against the Greek Logicians*, 11, with minor alterations.

[22] Daston and Galison, *Objectivity*, 210.

[23] Daston and Galison, *Objectivity*, 198.

arbitrary or untrustworthy. The scientist who was initiated into the community literally became the carrier of a lived, imparted, and mediated science of which he was one subject among many subjects over time.[24]

This situation as a whole presents us with a picture in which both the science and the experience that come to reside within a given scientist are objective, in our sense of the word. But for our historical actors, as we have explained, the dividing line between scientist and non-scientist was not a matter of objective versus subjective but one of expertise or initiation: the scientist could relate to the world scientifically, as opposed to the non-scientist, who could not. This holds true not only for objects of experience in the external physical world—such as minerals, plants, or animals—but also for objects that the scientist could investigate within himself, particularly his soul and its workings, which included dreams and introspections (as will become clear in Jules Janssens's chapter in this volume).

Revisiting Premodern Islamic Science and Experience, then, advocates a conceptual shift towards a historiography of experience and science that starts from the subject of the scientist as the substance within which, for our premodern actors themselves, science could naturally arise. This shift opens up the possibility of reconsidering the very demarcation between experience and science. By making scientists themselves the foundation of experience and science, we can begin to contemplate their cognitive being as a whole, thus both reshaping our view of the intellectual tradition of premodern Islamic science and enriching our appreciation of the complex dynamics that have driven scientific inquiry in other parts of the world and at other times.[25]

---

[24] Interesting cases may arise when epistemic values shift, as was the case when the post-Avicennan *falsafa* tradition gained considerable influence over the *kalām* tradition. In this respect, we may consider al-Ghazālī's statement on *ʿilm al-kalām* from his autobiography: theology is a science whose aim is only to defend the principles of Islam, but its method is not demonstrative, and thus of little use to one who seeks certain knowledge and admits only logically necessary truths and demonstrative arguments. See al-Ghazālī, *al-Munqidh*, 175. Al-Ghazālī thus broke from his community of theologians by considering their methods as untrustworthy. He praised their efforts, but this science could not teach him anything. For a different perspective on the *mutakallimūn* and their scientific aspirations, see, for instance, Dhanani, *The Physical Theory of Kalām*.

[25] In *Islamic Science and the Making of the European Renaissance* (2007), George Saliba argued for "a new historiography that could better explain the scientific developments [...] in the intellectual history of Islamic civilizations" (vii). His new historiography meant going "back to the primary sources, both historical and scientific," and trying "to re-read them without biases of any ideological narrative, as much as possible" (71–72). Although Saliba focused on a different aspect of the history of premodern Islamic science, he objected—just as we do here—that these historiographies often reflect the ideas of their modern-day authors as opposed to doing justice to the ways in which historical actors thought.

1.5 The Ground Plan

Experience could give rise to cognitive practices including listening, reading, observing, experimenting, reasoning, and judging. It was employed to investigate scientific objects ranging from bodily diseases, to the soul, to the heavens, to God. For Abū Maʿshar (d. 272/886), as we learn from **Charles Burnett**'s contribution, experience was a necessary part of the science of astrology. Astrology required experience in the empirical sense: perceptual data on the world, observations of heavenly bodies, and the influences of celestial bodies on the terrestrial realm were all used to make predictions about peoples' lives. Most importantly, these functional kinds of experience did not exist outside the astrologer, the subject of the science of astrology—they were embodied and ensouled practices of science, and accordingly were only partially observable for the uninitiated. Indeed, the guiding thread of Burnett's story is that even though the experience relevant for astrology is shared by all people, including laymen and practitioners of other arts, that experience is rendered scientifically relevant—part of the science of astrology—only by the specific epistemic expertise of the astrologer. The astrologer is the one who has learned to experience, or "read," heavenly phenomena in the relevant way—as signs that are mediated by analogical reasoning and by experience transmitted from authorities—and who can structure and order his knowledge. There is thus a crucial qualitative difference between the experience shared by all and the experience reserved for the community of astrologers.

As a source of scientific knowledge, experience could embrace both the scientist's outer senses and the inner senses of his soul ("soul," again, understood as our historical actors understood it), and could even cover different states of his consciousness, including both wakefulness and sleep. This wide spectrum of scientific experience emerges from **Jules Janssens**'s contribution, which presents Ibn Sīnā's discussion of veridical dreams. As Janssens explains, Ibn Sīnā regarded veridical dreams not only as an object of study in his science of the soul, but also, or even primarily, as a path to scientific knowledge about future contingencies and a tool for medical diagnosis. Perhaps somewhat surprisingly, Ibn Sīnā associated dream experience with *tajriba*, a concept that has received much scholarly attention in the past and has been translated as "methodic experience" or "experimentation."[26] However, in the context of dream experience, this is not a full-fledged method of collating and inferencing, but the dreamer's personal, sudden undergoing of a state of awareness with unique content. Far from denigrating that personal state of awareness as something merely subjective and thus unscientific, as might be done today, Ibn Sīnā incorporated such dreams into his scientific practice. And when it came to dream interpretation, Janssens tells us, the dreamer's previous ordinary experiences were combined with rational conjectures to build the methodological foundation for generating scientific knowledge about the content of dreams. For Ibn Sīnā, in other words, both the individual's experience

[26] For more on these issues, see Krause with Auxent and Weil, "Introduction: Making Sense of Nature in the Premodern World."

while dreaming and the rationally governed experience of the trained philosopher, both in their full phenomenological breadth, were required in order to attain scientific knowledge.

The qualitative difference between the experience shared by all and the experience reserved for the community of astrologers (or other communities of scientists), as Burnett's and Janssens's contributions highlight, constitutes a very particular dimension of the epistemic internalization of experience, whereby a small group of people claims to possess a more intimate relation with its object of knowledge than any other group in society. It is at this point that experience, as internalized and objectified by a specific group of people, became an object of social negotiation and battle. In the course of some such battles scientific experience—in the wide sense we outlined above—came to be granted or denied legitimacy as a source of scientific knowledge. In the most politicized circumstances, experience could even be used as a weapon in the establishment of social superiority and hegemony.

This is the story that emerges from **Nimrod Hurvitz**'s chapter on the caliphal *miḥna*, or inquisition, pursued from 833 to 848 CE. Two groups of practitioners of theology, the speculative theologians (*mutakallimūn*) and the traditionists (*muḥaddithūn*), fought fiercely over the question whether experience, in conjunction with human reason, can be a legitimate source of knowledge of God. The *mutakallimūn* affirmed this. The traditionists denied it, and argued that revelation was the only permissible source of such knowledge. From these two diametrically opposed epistemic norms arose distinct epistemic practices, of which Hurvitz mentions three: rational investigation as opposed to the following of authorities (*taqlīd*), metaphorical versus literal interpretation of the Qur'an, and the approval or rejection of debate. Carried out over generations and solidified within two opposed groups of practitioners of theological science, these practices gave rise to a social split and rival visions of social hierarchies in which one group would rule over the other—justified in each case by the expertise in the science of theology inherited and practiced within their own ranks.

In all three main chapters, it becomes clear that scientific experience was credited with internalized objectivity by its premodern Islamic possessors, the historical actors themselves. **Jon McGinnis** reflects upon this finding in his epilogue to the book. In a detailed engagement with the three main chapters and his own elucidating examples of historical reflection about the embeddedness of science and scientific experience in "epistemic cultures," McGinnis underscores the importance of our volume's central vantage point: the subject-rootedness of experience. He contends that whether some present-day historians of science endorse or oppose this historical given is not a question relevant to our historical research, and perhaps should not be for any historical research. The question, instead, is—or should be—whether our histories do justice to the scope and role of any given epistemic category in the place and at the time when it constituted science.

We have endeavored to take the first steps towards telling that kind of story for the epistemic category of scientific experience, reorienting the questions asked of premodern Islamic science away from what kinds of *observations* and *experiments* begot premodern science to what kinds of *experiences* begot premodern science, and how precisely they did so within the subject of science: the scientist. We hope our new architecture will prove to be equally fruitful for reexamining the epistemic category of experience in those Islamic sciences that are usually regarded as relying on empirical data, such as astronomy and medicine.

Bibliography

Akasoy, Anna A., and Alexander Fidora. 2016. The Structure and Methods of the Sciences. In *The Routledge Companion to Islamic Philosophy*, ed. Richard C. Taylor and Luis Xavier López-Farjeat, 105–114. London: Routledge.

Allen, Barry. 2021. *Empiricisms: Experience and Experiment from Antiquity to the Anthropocene.* Oxford: Oxford University Press.

Anawati, Georges C. 1981. Ishām Ibn Sīnā fī taqaddum al-ʿulūm. *Al-Turāth Al-ʿarabī* 2 (5–6): 16–42.

Basalla, George, ed. 1968. *The Rise of Modern Science: External or Internal Factors?* Lexington, MA: D. C. Heath.

Briffault, Robert. 1919. *The Making of Humanity.* London: Allen & Unwin.

Butterfield, Herbert. 1965. *The Origins of Modern Science.* New York: The Free Press.

Brentjes, Sonja. 2018. *Teaching and Learning the Sciences in Islamicate Societies, 800–1700.* Turnhout: Brepols.

Bynum, W.F., E.J. Browne, and Roy Porter 1981. *Dictionary of the History of Science.* London: Macmillan.

Cassirer, Ernst. [1910] 1923. *Substance and Function, and Einstein's Theory of Relativity*, translated by William Curtis Swabey and Martha Collins Swabey. Chicago: Open Court.

Clagett, Marshall. 1959. *The Science of Mechanics in the Middle Ages.* Madison: University of Wisconsin Press.

Crombie, Alistair Cameron. 1953. *Robert Grosseteste and the Origins of Experimental Science, 1100–1700.* Oxford: Clarendon Press.

Dales, Richard C. 1961. Robert Grosseteste's Scientific Works. *Isis* 52 (3): 381–402.

Daston, Lorraine. 2017. The History of Science and the History of Knowledge. *KNOW: A Journal on the Formation of Knowledge* 1 (1): 131–154.

Daston, Lorraine, and Peter Galison. 2010. *Objectivity.* New York: Zone Books.

Daston, Lorraine, and Katharine Park. 1998. *Wonders and the Order of Nature 1150–1750.* Princeton, NJ: Princeton University Press.

Daston, Lorraine, and H. Otto Sibum, eds. 2003. Scientific Personae and Their Histories. Special issue, *Science in Context* 16 (1–2).

Dear, Peter. 2006. The Meanings of Experience. In *The Cambridge History of Science*, ed. Katharine Park and Lorraine Daston, 106–131. Cambridge: Cambridge University Press.

Dhanani, Alnoor. 1994. *The Physical Theory of Kalām: Atoms, Space, and Void in Basrian Muʿtazilī Cosmology.* Leiden: Brill, 1994.

Dijksterhuis, E.J. 1961. *The Mechanization of the World Picture: Pythagoras to Newton.* Translated by C. Dikshoorn. Oxford: Clarendon Press.

Duhem, Pierre Maurice Marie. 1903. *L'évolution de la mécanique.* Paris: A. Hermann.

Duhem, Pierre. 1970. *To Save the Phenomena: An Essay on the Idea of Physical Theory from Plato to Galileo.* Translated by Edmund Doland and Chaninah Maschler with an introduction by Stanley L. Jaki. Chicago: University of Chicago Press.

Dupré, Sven, and Geert Somsen. 2019. The History of Knowledge and the Future of Knowledge Societies. *Berichte Zur Wissenschaftsgeschichte* 42 (2–3): 186–199.

Eamon, William. 1996. *Science and the Secrets of Nature: Books of Secrets in Medieval and Early Modern Culture.* Princeton, NJ: Princeton University Press.

Gal, Ofer. 2021. *The Origins of Modern Science: From Antiquity to the Scientific Revolution.* Cambridge: Cambridge University Press.
al-Ghazālī, Abū Hāmid. 2001. *Al-Munqidh min al-ḍalāl*, ed. Muḥammad Muḥammad Abū Layla and Nūrshīf ʿAbd al-Raḥmān Rifʿat. Washington, DC: Council for Research in Values and Philosophy.
Grant, Edward. 1996. *The Foundations of Modern Science in the Middle Ages: Their Religious, Institutional and Intellectual Contexts.* Cambridge: Cambridge University Press.
Gutas, Dimitri. 2002. Certainty, Doubt, and Error: Comments on the Epistemological Foundations of Medieval Arabic Science. *Early Science and Medicine* 7 (3): 276–289.
Gutas, Dimitri. 2012. The Empiricism of Avicenna. *Oriens* 40 (2): 391–436.
Hallaq, Wael B., trans. 1983. *Ibn Taymiyya against the Greek Logicians.* Oxford: Clarendon Press.
Hallaq, Wael B. 1993. Introduction. In *Ibn Taymiyya Against the Greek Logicians*, translated with an introduction and notes by Wael B. Hallaq, xi–lviii. Oxford: Clarendon Press.
Hanson, Norwood Russell. 1958. *Patterns of Discovery: An Inquiry into the Conceptual Foundations of Science.* Cambridge: Cambridge University Press.
Hossfeld, Paul. 1983. *Albertus Magnus als Naturphilosoph und Naturwissenschaftler.* Bonn: Albertus-Magnus-Institut.
Iqbal, Muhammad. [1930] 2013. *The Reconstruction of Religious Thought in Islam.* Stanford, CA: Stanford University Press.
Jacquart, Danielle. 2017. Die Medizin als Wissenschaftsdisziplin und ihre Themen. In *Die Philosophie des Mittelalters*, vol. 4: *13. Jahrhundert*, ed. Alexander Brungs, Vilem Mudroch, and Peter Schulthess, 1595–1612. Basel: Schwabe.
Janssens, Jules. 2004. 'Experience' (*tajriba*) in Classical Arabic Philosophy (al-Fārābī–Avicenna). *Quaestio* 4 (1): 45–62.
Karimullah, Kamran I. 2017. Avicenna and Galen, Philosophy and Medicine: Contextualising Discussions of Medical Experience in Medieval Islamic Physicians and Philosophers. *Oriens* 45 (1–2): 105–149.
Klein-Franke, Felix. 1982. *Vorlesungen über die Medizin im Islam.* Wiesbaden: Steiner.
Krause, Katja, with Maria Auxent, and Dror Weil. 2022. Introduction: Making Sense of Nature in the Premodern World. In *Premodern Experience of the Natural World in Translation*, ed. Katja Krause, Maria Auxent, and Dror Weil, 7–19. New York: Routledge.
Kusukawa, Sachiko, and Ian Maclean. 2006. *Transmitting Knowledge: Words, Images, and Instruments in Early Modern Europe.* Oxford: Oxford University Press.
Lakatos, Imre. 1971. History of Science and Its Rational Reconstructions. *Boston Studies in the Philosophy of Science* 8: 91–136.
Leong, Elaine. 2018. *Recipes and Everyday Knowledge: Medicine, Science, and the Household in Early Modern England.* Chicago: University of Chicago Press.
Mach, Ernst. [1905] 1914. *The Analysis of Sensations, and the Relation of the Physical to the Psychical.* translated by Cora May Williams and Sydney Waterlow. Chicago: Open Court.
Maier, Anneliese. 1966. *Die Vorläufer Galileis im 14. Jahrhundert.* Rome: Storia e Letteratura.
Maier, Anneliese. 1958. *Zwischen Philosophie und Mechanik: Studien zur Naturphilosophie der Spätscholastik.* Rome: Storia e Letteratura.
McEvoy, J.J. 1982. *The Philosophy of Robert Grosseteste.* Oxford: Clarendon Press.
McGinnis, Jon. 2008. Avicenna's Naturalized Epistemology and Scientific Method. In *The Unity of Science in the Arabic Tradition: Science, Logic, Epistemology and Their Interactions*, ed. Shahid Rahman, Tony Street, and Hassan Tahiri, 129–152. Dordrecht: Springer.
McGinnis, Jon. 2003. Scientific Methodologies in Medieval Islam. *Journal of the History of Philosophy* 43 (3): 307–327.
Pomata, Gianna, and Nancy G. Siraisi, eds. 2005. *Historia: Empiricism and Erudition in Early Modern Europe.* Cambridge, MA: MIT Press.
Rankin, Alisha. 2021. *The Poison Trials: Wonder Drugs, Experiment, and the Battle for Authority in Renaissance Science.* Chicago: University of Chicago Press.

Renn, Jürgen. 2015. The History of Science and the Globalization of Knowledge. In *Relocating the History of Science*, ed. Theodore Arabatzis, Jürgen. Renn, and Ana Simões, 241–252. Cham: Springer.
Rosenthal, Franz. 1975. *The Classical Heritage in Islam*. London: Routledge & Kegan Paul.
Rosenthal, Franz. 1990. *Science and Medicine in Islam: A Collection of Essays*. Aldershot, UK: Variorum.
Saliba, George. 2007. *Islamic Science and the Making of the European Renaissance*. Cambridge, MA: MIT Press.
Saliba, George. 1987. Theory and Observation in Islamic Astronomy: The Work of Ibn al-Shāṭir of Damascus. *Journal for the History of Astronomy* 18: 35–43.
Schulthess, Peter. 2017. Methode in der Naturphilosophie. In *Die Philosophie des Mittelalters*, vol. 4: *13. Jahrhundert*, ed. Alexander Brungs, Vilem Mudroch, and Peter Schulthess, 1363–1366. Basel: Schwabe.
Shank, Michael H., and David C. Lindberg. 2013. Introduction. In *The Cambridge History of Science*, vol. 2: *Medieval Science*, ed. David C. Lindberg and Michael H. Shank, 1–26. Cambridge: Cambridge University Press.
Shapin, Steven. 2010. *Never Pure: Historical Studies of Science as If It Was Produced by People with Bodies, Situated in Time, Space, Culture, and Society, and Struggling for Credibility and Authority*. Baltimore, MD: Johns Hopkins University Press.
Shapin, Steven. 1992. Discipline and Bounding: The History and Sociology of Science as Seen Through the Externalism–Internalism Debate. *History of Science* 30: 333–369.
Shapin, Steven, and Simon Schaffer. 1985. *Leviathan and the Air-Pump: Hobbes, Boyle, and the Experimental Life*. Princeton, NJ: Princeton University Press.
Thorndike, Lynn. 1929–1958. *A History of Magic and Experimental Science*. 8 vols. London: Macmillan.
Thorndike, Lynn. 1914. Roger Bacon and Experimental Method in the Middle Ages. *The Philosophical Review* 23 (3): 271–298.
Truitt, E.R. 2015. *Medieval Robots: Mechanism, Magic, Nature, and Art*. Philadelphia: University of Pennsylvania Press.
Werrett, Simon. 2019. *Thrifty Science: Making the Most of Materials in the History of Experiment*. Chicago: University of Chicago Press.
Young, Mark Thomas. 2019. Experimentalist as Spectator: The Phenomenology of Early Modern Experimentalism. In *The Past, Present, and Future of Integrated History and Philosophy of Science*, ed. Emily Herring, Kevin Jones, Konstantin Kiprijanov, and Laura Sellers, 133–149. London: Routledge.

Hannah C. Erlwein holds a Ph.D. in Islamic intellectual history and specializes in the premodern Islamic sciences of *kalām* (theology) and *falsafa* (philosophy). Her work currently focuses on debates about the nature and attainment of scientific knowledge in these two sciences.

Katja Krause is a historian of science and philosophy at Technische Universität Berlin and leads the Max Planck Research Group "Experience in the Premodern Sciences of Soul and Body, ca. 800–1650" at the Max Planck Institute for the History of Science, Berlin.

Chapter 2
"Obvious, Clear, and in Front of Our Eyes": Defending the Science of Astrology by Means of Experience

Charles Burnett

Abstract In premodern Islamic astrology, the demarcation between knowledge gained by experience and theoretical knowledge is difficult to discern. Astrology is often described as a *ṣināʿa* (craft or profession) rather than an *ʿilm* (science) and is therefore comparable to other *ṣināʿāt* (crafts)—such as animal husbandry, midwifery, and navigation—in which skill is combined with experience (*tajriba, experimentum*). Common to all these crafts is also the experience of one's teachers and predecessors. What is added by the astrologers, along with the doctors, is the structured ordering of such experiences in written form and the attempt to explain the experienced phenomena on a theoretical basis. Both astrologers and doctors were subject to examinations by authorities, with the aim of proving that they had scientific knowledge that could be applied in practice. This study focuses on the contribution to the discussion of theory and practice of the *Great Introduction to Astrology* by Abū Maʿshar (Albumasar, 787–886 CE).

Keywords Premodern Islamic astrology · Experience · *Tajriba* · *Experimentum* · Abū Maʿshar (Albumasar)

With the emergence of Islam in the seventh century CE, numerous Islamic sciences began to arise. From the moment these sciences entered the stage, their practitioners started to debate the legitimacy of their undertakings and the epistemic status of the various disciplines. A famous example is the dispute between speculative theologians (*mutakallimūn*) and traditionists (*muḥaddithūn*) over the science of *kalām*.[1] The practitioners of *kalām* argued that religious dogmas must be defended by means of rational arguments and that this was the task they were undertaking, but from

[1] The *muḥaddithūn*, here translated as "traditionists," are also known as "Traditionalists": see Hurvitz in this volume.

C. Burnett (✉)
University of London, London, UK
e-mail: charles.burnett@sas.ac.uk

H. C. Erlwein and K. Krause, *Revisiting Premodern Islamic Science and Experience*, SpringerBriefs in History of Science and Technology,
https://doi.org/10.1007/978-3-031-76085-3_2

the perspective of the traditionists, *kalām* was irreligious because it contradicted authoritative religious texts. As such, the traditionists argued, its practitioners upheld flawed epistemic norms and values, so that *kalām* did not merit being called a science. Nimrod Hurvitz, in this volume, analyzes this dispute in greater detail.

A similar case is that of astrology. In his *Book on the Great Introduction to Astrology*, Abū Maʿshar (Albumasar, 787–886 CE) goes to great lengths to defend the legitimacy of astrology as a science against its (unnamed) detractors. They present various types of arguments:

> I saw some people disputing about the essence of astrology, and some said that the power of the movements of the stars has no effect in this world at all, while others said that it has an effect on genera, species, and the four elements, but on nothing else. | Some said that it has an effect on the changes and alterations of times [i.e., seasons] only; others said, while rejecting different opinions, that it has an effect on everything in this world, this being the opinion of the masters of the art of the stars. I did not see any of these give clear proofs for their statements, <of the kind> that intelligent people can accept.[2]

Abū Maʿshar goes on to say that "many people think that astrology is a thing that people find merely by conjecture and guesswork, and that it does not have a sound basis on which to operate or to use as a standard."[3] Abū Maʿshar seeks to refute this and the remaining objections in a section headed "On the existence of the science of astrology."[4] There, he characterizes astrology as a systematic body of knowledge that follows certain scientific norms and values. At the heart of his defense of astrology as a science stands experience: the experiences of different individuals and groups are evidence of the truth of astrology's central claims. In this chapter, I analyze Abū Maʿshar's defense of astrology as a science, with particular attention to his invocation of experience.

2.1 Astrology as a Science

Astrology is often described in Arabic as *ṣināʿat aḥkām al-nujūm*, "the art of the judgments of the stars"; *ṣināʿa* is equivalent to the Greek *tekhnē* and Latin *ars* or *magisterium*.[5] It is a noun formed from the root *ṣ-n-ʿ*, which gives the verb *ṣanaʿ*, "to make." This verb is commonly found in recipe books, including those that present

---

[2] Abū Maʿshar, *The Great Introduction to Astrology*, 1, 1.7a–b, ed. and trans. Yamamoto and Burnett (hereafter *Great Introduction*, translation slightly modified throughout), vol. 1, 45. Abū Maʿshar goes into more detail a little later, listing and refuting ten categories of lay objections to astrology. Ibid., 1, 5, vol. 1, 106–149.

[3] *Great Introduction*, 1, 1.9a, vol. 1, 46–47.

[4] *Great Introduction*, 1, 2, vol. 1, 52–53.

[5] Al-Qabīṣī refers to astrology as *ṣināʿat aḥkām al-nujūm* at the beginning of his popular *Introduction to Astrology*, 18. This is translated *magisterium iudiciorum astrorum* in the even more popular Latin version (ibid., 225). The same term is used regularly in the title of this book as well as in the same author's *Kitāb fī ithbāt ṣināʿat aḥkām al-nujūm* (The Book on the Confirmation of the Craft of Astrology) and *Risāla fī imtiḥān al-munajjimīn* (The Treatise on the Testing of the Astrologers).

recipes for magical practices.[6] A *ṣināʿa* is a practical activity that might be expected to rely on, or develop through, experience. However, the ordered structure of theoretical knowledge that underlies the practice of astrology merits its description as an *ʿilm*—a science—and *ʿilm aḥkām al-nujūm* also occurs. The latter denomination is found in the title of Abū Maʿshar's *Book on the Great Introduction to Astrology*, the *Kitāb al-mudkhal al-kabīr ilā ʿilm aḥkām al-nujūm.*[7]

Abū Maʿshar gives some idea of what is involved in such a science or craft in his definition of astrology and summary of its parts. The definition is as follows: "Astrology is the knowledge [*ʿilm*[8]] of what the power of the movements of the planets at a specific time indicates for that time and for a specified future time."[9] The summary of the parts of astrology and the attitude of the astrologer, too, indicates its scientific character: "Astrology has a starting-point [*ibtidāʾ*, *initium*], a root [*aṣl*, *radix*], a branch [*farʿ*, *ramus*], a proof [*burhān*, *auctoritas*], a fruit [*thamr*, *fructus*], and a completion [*tamām*, *perfectio*]."[10] Abū Maʿshar elaborates as follows:

> The starting-point for making judgements that are passionately desired [*maʿshūq*, *admirabilis*] is outstanding knowledge of the science of things coming-to-be [*al-ashyāʾ al-kāʾināt*, *res que fiunt*] and interest [*ʿināya*, *sollicitudo*] in them.
>
> Its root is the knowledge of the quality and quantity of the movements of the celestial bodies.
>
> The branch of this knowledge is to judge by them matters existing in this changeable world [*al-umūr al-kāʾināt fī hādhihi al-ʿālam al-mutaghayyir*, *res que fiunt in hoc mundo corruptibiles*].[11]
>
> The proof of the judgements is the correctness [*ṣuwāb*, *veritas*] which comes about by prediction from the conditions of the planets and their action on the thing about which information is sought, among the things which will happen.
>
> The acquisition of this science only comes about with difficulty and labor, and correctness concerning existing things by opinion [*raʾy*, *arbitrium*] and estimation [*takhmīn*, *ratio*] may be available only to a special kind of person at certain times. [...]
>
> The fruit is the correctness, benefit and usefulness coming from it [...].

---

[6] The use of *ṣanaʿ* is fully explored in Burnett, "Practical Magic."

[7] Abū Maʿshar's preferred title is *ʿilm aḥkām al-nujūm* (*scientia iudiciorum astrorum*; Abū Maʿshar, *Great Introduction*, vol. 1, 30–32), though within his text, *ṣināʿat al-nujūm* (*magisterium astrorum*) is used almost as frequently as *ʿilm al-nujūm* (*scientia astrorum*: ibid., vol. 2, 341), and in the opening protocol the two terms are combined: Abū Maʿshar "composed this book on the art [*ṣināʿa*, *magisterium*] of the introduction to the science [*ʿilm*, *scientia*] of astrology" (ibid., 1, 1.4, vol. 1, 43).

[8] Note that *ʿilm* is translated both as "science" and as "knowledge," according to context. When it is referring more closely to the subject matter at hand, it is a "science"; when it is referring to the attitude of mind of the scientist, it is "knowledge."

[9] Abū Maʿshar, *Great Introduction*, 3, 2.3, vol. 1, 232–233.

[10] Abū Maʿshar, *Great Introduction*, 3, 2.14a, vol. 1, 240–241. In the following, the Arabic terms and phrases will be presented as they occur in Abū Maʿshar, *Great Introduction*, while the Latin equivalents are from the literal translation by John of Seville as they occur in Abū Maʿshar al-Balkhī, *Kitāb al-Mudkhal al-kabīr ilā ʿilm aḥkām al-nujūm*, ed. Lemay, vol. 5. Occasionally, reference will be made to the more paraphrastic translation of Hermann of Carinthia (ibid., vol. 8).

[11] John of Seville transfers "changeable" from the "world" to the "things."

The benefit through correctness is its completion.[12]

Abū Maʿshar's presentation of astrology as a structured, systematic body of knowledge thus proceeds by identifying a starting point, root, and branch and introducing the ideas of knowledge, proof, and correctness. As a former *ḥadīth* scholar, he probably owed these terms to the jurisprudential tradition, in which root, branch, and proof are crucial.[13] At the same time, however, his list does not present astrology as an abstract body of systematic knowledge alone. He also stresses the role of the individuals involved in it, emphasizing that the processes he describes are oriented on the attitude of the participants. The client should "passionately desire" the outcome of his consultation, and the astrologer should not only be knowledgeable, but also have a deep interest in what he is doing. He must have a special gift to overcome the difficulties of obtaining results. Knowledge of the material, the right attitude, and a certain amount of intuition are necessary for making the right judgments. Abū Maʿshar is here stating in different terms, and as a series of definitions rather than continuous narrative, the epistemology of astrology that was set out in the second century CE in Ptolemy's *Tetrabiblos*, Book 1, Chap. 1.[14]

2.2 Reading the Signs

Astrology is very much rooted in the world around us and in the specific time in which we inhabit that world. A major argument is that ordinary people know by experience about hidden things or future events.[15] People have a natural curiosity about what is going to happen—and this curiosity has a practical side, since, once the future is known, one can take precautions and make plans to meet any eventualities. One can even feel the pleasure of excitement about a future event, for example, enjoying in advance the music one is going to hear. In this sense, on Abū Maʿshar's account, astrology is a science of direct relevance and benefit not only to its practitioners, but also to the wider public.

As well as noting the role of curiosity and experience in the general public's recourse to astrology, Abū Maʿshar also invokes the experience of particular groups in society as a way of gaining knowledge about concealed or future events. People in different professions can predict what will happen, or reveal what is hidden, with remarkable accuracy because of their own experience (*tajriba*, *experimentum*) and

[12] Abū Maʿshar, *Great Introduction*, 3, 2.14a–b, vol. 1, 240–243.

[13] Vadet, "Une défense de l'astrologie," 141–143, gives a detailed explanation of how this process follows the method of *qiyās* practiced by Arabic theoreticians.

[14] Note that Ptolemy emphasizes the difficulty of the subject matter of astrology, but asserts that it is nevertheless capable of being grasped in a philosophical way.

[15] The term *tajriba*, always translated by John of Seville as *experimentum* and here as "experience," occurs twenty-seven times in Abū Maʿshar's *Great Introduction*.

that of previous generations in their professions.[16] Abū Maʿshar gives as examples farmers, herdsmen, sailors, and shipmasters. All these professions observe the rising of specific constellations in order to determine the best time to sow, to mate animals, and to set out to sea.

But there are other professions in which the stars are not observed. For example, midwives know purely from their experiences (*tajārib*, *experimenta*) whether or not a woman is pregnant and whether the child will be a girl or a boy; they also know from the birth of the firstborn whether or not the woman will have another child and, if so, how many. Midwives rarely make mistakes, which is due to their own long experience and the information they have collected from "their predecessors who had experienced these things over a long period of time."[17] The examples Abū Maʿshar gives are very specific:

> If the tips of the breast (her nipples) have increased in size and changed color, the woman is pregnant.
>
> If her eyes are deeply sunken in their sockets, her eyelids have become slack, and her gaze is sharp, her pupils bright and full, and the whites of her eyes viscous, she is pregnant.
>
> If her belly is full, round and firm, the fetus is male; if she has a flaccid belly, it is female.
>
> If her nipples have turned black, the fetus is female; if they have turned red, male.
>
> If the breast milk is viscous, the fetus is male; if it is thin, it is female.
>
> If the breast milk coagulates like a pearl bead when placed on a mirror in the sun, again the fetus is male; if it spreads out, it is female.
>
> A crown of thin hair on the newborn means that the next child will be male.
>
> A double crown means that two males will be born.
>
> If the amniotic sac remains intact, this is a good sign.
>
> The lumps and knots on the umbilical cord indicate the number of children to be born subsequently.[18]

In all these cases, experience enables the professional to learn to detect signs in one (natural) thing which indicate something (specific) about another; they are signs from which judgments and predictions can be made. Abū Maʿshar's argument is that if people of such a lowly and uneducated status as midwives can predict successfully (at least within their profession) from their own experiences and those of their predecessors, how much better should an astrologer predict, given that he can use both his

---

[16] This is discussed in the First Part, Chaps. 2 ("On the existence of the science of astrology," vol. 1, 52–79) and 6 ("On the benefit of the science of astrology," vol. 1, 150–175) of Abū Maʿshar's *Great Introduction*. Again, it follows Ptolemy's *Tetrabiblos*, in which the second and third chapters of the first book are on the attainability and usefulness of astrology respectively.

[17] Abū Maʿshar, *Great Introduction*, 1, 2.18a, vol. 1, 62–63. Hermann of Carinthia begins his paraphrase of this paragraph with the almost proverbial phrase *experimenta fidem gerunt*, "experiences provide confidence" (Abū Maʿshar al-Balkhī, *Kitāb al-Mudkhal al-kabīr ilā ʿilm aḥkām al-nujūm*, 8, 6.162–6.163). The equivalent passage in John of Seville's translation, missing in Lemay's edition, is provided in the Appendix below.

[18] Abū Maʿshar, *Great Introduction*, 1, 2.18b–21, vol. 1, 63–65. See Appendix for part of this passage in Latin.

own experience and that of generations of other astrologers who have written down their experiences in books.

There is a more fundamental distinction as well: "The profession of medicine, along with the other professions [e.g., midwifery], is terrestrial; its subject is bodies and individuals which fade away and alter and which receive increase and decrease, coming-to-be and passing-away," whereas astrology's subject is "the stars, which do not alter and are not subject to coming-to-be and passing-away, for as long as God wills.[19] The astrologer observes the most exalted, stable, and rational beings in the universe; therefore, his predictions have the potential to be more reliable and more universal than those of any other profession.

How do we know that predictions based on the movement of the stars really are reliable? I have said that experience informs the arts and sciences when signs are detected in one matter that indicate something about another matter. In this case, the signs are the movements of the planets, which—though in part "obvious" to everyone—are ultimately only fully interpretable to the skilled astrologer, since he can apply the appropriate scientific practice of analogous reasoning. Abū Maʿshar writes:

> What is obvious of the science of the universe is discovered by observation [*mawjūd bi-l-ʿiyān, visu percipitur*]. To what is not found by observation, analogy is applied [*qiyās*, translated into Latin as *experimentum* by John of Seville and as *ratio* by Hermann of Carinthia], which compels one to accept it, because the indications and the proofs of it are part of the obvious and clear causes on which there is agreement, taken from the sciences of arithmetic, geometry, and land measuring, with which no doubt can be mingled.[20]

From these astronomical observations is derived astrological information, that is,

> knowledge of the nature of every planet and every sphere, the property of their indications, and what arises and happens as a result of the powers of their different movements and their natural imprint on this world, which is under the sphere of the moon.[21]

Many conditions and events, Abū Maʿshar says, are "obvious, clear, and in front of our eyes." They are signs the astrologer can read by way of observation and experience. As for "what is not obvious," this "is inferred by means of clear analogies [*al-qiyāsāt al-wāḍiḥa, patens experimentum*] and from natural science."[22] Abū Maʿshar proceeds to give many examples of both the obvious and the derived knowledge. By observation, one can see that the changes of the seasons follow the transfer of the

---

[19] Abū Maʿshar, *Great Introduction*, 1, 2.26a and 1, 2.25c, vol. 1, 70–71. For a detailed assessment of the contrast between the predictions of doctors and those of astrologers as a result of their respective experience, see Burnett, "Doctors versus Astrologers."

[20] Abū Maʿshar, *Great Introduction*, 1, 2.3c, vol. 1, 52–55.

[21] Abū Maʿshar, *Great Introduction*, 1, 2.4, vol. 1, 54–55.

[22] Abū Maʿshar, *Great Introduction*, 1, 2.5, vol. 1, 54–55. Note that Hermann of Carinthia emphasizes the sensual aspect of the experiences, and their testing by repetition, in his translation of this passage, which nicely combines experience and theory: "partim crebris quibusdam sensibilibus experimentis, partim naturali speculatione" (partly by certain frequent sensible experiences, partly by natural speculation) (Abū Maʿshar al-Balkhī, *Kitāb al-Mudkhal al-kabīr ilā ʿilm aḥkām al-nujūm*, 8, part 4, 1, 1.95–1.96).

sun from one quadrant of the sphere to another; changes of temperature and humors follow the daily path of the sun through the sky; the movements of certain plants follow the sun; and the movement of the sea follows the waxing and waning of the moon. By analogy, the effect of the other planets must account for the differences of the seasons from year to year.[23]

As Abū Maʿshar shows at length, therefore, the justification of astrology arises largely from experience and direct observation of nature, whether present-day experiences or experiences of the past. Among past experiences are those of the ancient observers of the heavens. In Hermann of Carinthia's reading of Abū Maʿshar, for example, they observed that an image of the Virgin could be seen in the zodiac sign of Virgo, nourishing a child with her milk, which indicated that already well before the time of Christ, natural scientists were aware that Christ would be born.[24]

2.3 Scientific Experience

To an extent, then, astrology is adequately justified by people's universal experience. But how exactly does experience play a part in the practice of astrology as a science, and to what degree is this second kind of experience qualitatively particular to the trained, professional astrologer?

Abū Maʿshar identifies certain facets of the astrologer's experience that, in his view, make it scientific. First of these is the experience of the astrologer himself in his craft. Abū Maʿshar criticizes those who pretend to be astrologers when they have only a small knowledge of the craft. "The knowledge of these sciences," he says, "can only be acquired over a long period of time and with hard work."[25]

Second, to the objection that astrologers cannot experience twice in their lifetimes many of the phenomena on which they base their judgments (meaning that they cannot compare repeated events), Abū Maʿshar replies that they can rely on the observations and conclusions of generations of astrologers before them, whose knowledge they have inherited.[26]

Third, the astrologer can also use indirect experience in the form of the mathematical data (the observed movements of the heavenly bodies) compiled by arithmeticians, just as the doctor relies on the herbs and other medical materials collected by pharmacists.[27]

Abū Maʿshar is surprisingly silent about the astrologer's direct experience of his client. He mentions that there are subtle aspects about individuals—different gestures, different colors of skin, a sweet smell or bad odor—whose stellar origin one cannot

---

[23] Abū Maʿshar, *Great Introduction*, 1, 2.6a–2.10, vol. 1, 54–59.

[24] This is Hermann's interpretation of Abū Maʿshar, *Great Introduction*, 6, 1.31a, vol. 1, 572–573, in Hermann of Carinthia, *De essentiis*, ed. Burnett, 82.

[25] Abū Maʿshar, *Great Introduction*, 1, 5.41e, vol. 1, 146–147.

[26] Abū Maʿshar, *Great Introduction*, 1, 5.25a–5.27b, vol. 1, 128–131.

[27] Abū Maʿshar, *Great Introduction*, 1, 5.33b–c, vol. 1, 134–135.

judge: "The difference between them is found only by subtlety of sense-perception" (*bi-laṭīf al-ḥiss, per subtilitatem sensus*). This would be the sense-perception of the astrologer—a direct experience—but Abū Maʿshar goes on to say that such differences are not relevant to making an astrological judgment.[28]

The qualified astrologer, therefore, would seem to require iteration of experiences, or, at least, the ability to make judgments on new cases on the basis of his experience of similar cases in the past. It is his familiarity with different astral configurations and their effects that differentiates him from the layperson. But the experience relevant for astrology as science is not confined to the direct experiences of a given astrologer; it also encompasses the accumulated experiences of previous authorities. Experience here is an amalgam of direct, personal experiences and transmitted, authoritative experiences. These become relevant for astrology once mediated by theory. For instance, the astrologer needs to be familiar with the mathematical assumptions underlying the movements of heavenly bodies in order to be able to make his astrological predictions.

One of the aims of astrology mentioned at the beginning of this chapter, "correctness concerning existing things by opinion and estimation [*bi-l-raʾy wa-l-takhmīn, per arbitrium et rationem*], which is found in only a special kind of person,"[29] might additionally suggest some intuition beyond astrological know-how in the face of experienced things. The client, too, presumably contributes as the one who "passionately desires" a judgment on a topic. One finds more about the experience and attitude of the client in works of astrological elections (choices) or interrogations (questions), in which the client must be truthful and single-minded in what he chooses or asks a question about (he must have a strong *ḍamīr* or *intentio*) and must not ask a question designed to trip up or deceive the astrologer.[30] The experience of the astrological consultation, however, is not a concern of the *Great Introduction to Astrology.*

Considering the significance he attributes to direct and transmitted experience, it may be somewhat surprising that Abū Maʿshar does not address the topic of testing the success of astrological predictions. Neither does he compare these to other methods of prediction using a controlled experiment, although such a test might be implied in the aims and purpose of astrology I quoted earlier in this chapter: "The proof of the judgments is the correctness [*ṣuwāb*] which comes about." The material for a comparative test would be readily available in the *Great Introduction,* since Abū Maʿshar does give the astrological methods for asking several of the same questions concerning pregnancy and the fetus that, we have seen, were asked by the midwives.

This material is found in the sixteenth chapter of the Sixth Part of the *Great Introduction,* which concerns the zodiac signs indicating many children, twins, few children, and barrenness, but the chapter says only that "the signs having many children are Cancer, Scorpio, Pisces, and the latter half of Capricorn," and does not specify how many children there will be; the signs having few children are Aries,

---

[28] Abū Maʿshar, *Great Introduction*, 3, 2.8c, vol. 1, 236–267.

[29] Abū Maʿshar, *Great Introduction*, 3, 2.14b, vol. 1, 243.

[30] Mesehella (Māshāʾallāh), *De ratione inquirendi et modo*, MS Vienna, Österreichische Nationalbibliothek, clm 2428, fol. 1r, trans. Dykes in *The Book of the Nine Judges*, 43–44.

Taurus, Libra, Sagittarius, and Aquarius, but again no numbers are mentioned. Twins are indicated by the latter half of Capricorn, the bicorporeal signs (Gemini, Virgo, Sagittarius, and Pisces), and sometimes Aries and Libra, and barrenness by Gemini, Libra, Virgo, the beginning of Taurus, and sometimes Aquarius and the beginning of Capricorn.[31]

More information appears in the Eighth Part of the *Great Introduction*, which is devoted to the astrological lots that are cast for answering questions. They are divided amongst the twelve astrological houses (or divisions of the sky at the time of the question, starting from the ascendant), including self, possessions, brothers, and fathers, until we come to the subject of the fifth house, which is children. It is here that one casts lots to see whether one's client has children or not. The second lot indicates the time at which children are born, their number, and their gender. Here, Abū Maʿshar also supplies the means of calculating the number: the planetary lord of the place at which the lot arrives will indicate the number of children, according to the planet's years. For example, for Jupiter its "small years" are twelve, its middle years are forty-five-and-a-half, and its "great years" are seventy-nine—surely an enormous number of children![32] Probably, though, the meaning is only that small years indicate a small number of children, middle years, a middling number, and great years, many. This can be increased by an aspecting planet, which can add its own years to those of the first planet. The fifth lot is that by which one knows whether the fetus is male or female. If it falls in a masculine sign, the fetus is male; if in a feminine sign, female.[33]

To be sure, these are not very precise predictions, and they are not "tested" (*mujarrab*) either. In this case, the predictions of the midwife, with direct experience of delivering babies, are more precise. Perhaps Abū Maʿshar is so convinced of the "truth" of astrology and the reliability of its authorities that he does not see the necessity of testing the results.

It should also be added that, while astrology certainly uses experience, many astrological predictions are based only on analogy—the earthly scorpion follows the heavenly Scorpio, the earthly snake follows the heavenly Hydra, and so on.[34] The section of the *Great Introduction to Astrology* detailing the different indications of the signs of the zodiac is particularly apt at showing such analogies between heavenly signs and earthly phenomena. Signs indicating the voice are divided into those that are in a human shape, indicating a strong voice: Gemini, Virgo, and Libra. Those that have "half a voice" are in the shape of brute animals: Aries, Taurus, Leo, and Sagittarius. Signs having a weak voice are mixed animate beings (Capricorn and

---

[31] Abū Maʿshar, *Great Introduction*, 6, 16, vol. 1, 658–659.

[32] Abū Maʿshar, *Great Introduction*, 8, 4.23a, vol. 1, 866–867.

[33] Abū Maʿshar, *Great Introduction*, 8, 4.26, vol. 1, 870–871.

[34] This is stated most clearly in Aḥmad ibn Yūsuf's commentary of Pseudo-Ptolemy's *Centiloquium* in MS Tehran, Malik, 5924: "mithl an takūn al-ʿaqārib muṭīʿa li-l-ʿaqrab fī al-falak wa-l-ḥayyāt muṭīʿa li-ṣūrat al-shujāʿ fī l-falak" (Just as scorpions obey the scorpion in the heavenly sphere and snakes obey the Hydra in the heavenly sphere). This was later to be called the doctrine of signatures.

Aquarius), and those having no voice are Cancer, Scorpio, and Pisces.[35] Similarly, the signs indicating birds are the second and third decan of Capricorn because the Flying Eagle and the tail of the Hen (Cygnus) are there, whereas quadrupeds are indicated by Aries, Taurus, Leo, and the latter half of Sagittarius.[36] This prediction by analogy is clearly not based on experience, and it is questionable whether the analogies themselves have any basis in empirical testing. It seems more likely that Abū Maʿshar is employing certain theoretical assumptions, which, however, are not explained.

The idea of testing is absent in Abū Maʿshar when it comes to the expertise of the astrologer, but other historical actors indicate the existence of rules for astrology that can be tested. The work by al-Qabīṣī called the *Risāla fī imtihān al-munajjimīn* (Treatise on the Testing of the Astrologers) is one example. The treatise offers a list of thirty questions that can be asked of the prospective astrologer in order to test whether he is qualified to practice his profession. It is addressed to Sayf al-Dawla, the Ḥamdānid Emir of Aleppo from 945 to 967, and al-Qabīṣī explains its purpose as specifically being to single out the genuine astrologer from those who, out of either deceit or incompetence, only pretend to be astrologers (hence the full title of the work: The Treatise on the Testing of Those Who Call Themselves Astrologers).

To pass this selection process, it is necessary to have been trained by another astrologer. Al-Qabīṣī himself was a pupil of ʿAlī ibn Aḥmad al-ʿImrānī, who wrote an influential book on astrological elections and is described by Ibn al-Nadīm as having read Ptolemy's *Almagest* with al-Qabīṣī.[37] One could call this training and a successful outcome of the examination the astrologer's professional qualification.[38] The "complete astrologer," according to the *Imtihān al-munajjimīn*, is the one who is able to work out everything by himself, using his own intelligence. He makes his own astronomical tables by using direct observations.[39] The "testing" here is that of the astrologer, not of astrology itself, which is rather what one would find in books "on the confirmation of astrology" (*fī ithbāt ṣināʿat aḥkām al-nujūm*), one of which was written by al-Qabīṣī.[40]

Al-Qabīṣī reminds us that self-defined astrologers should not be taken at face value but should be tested as if they were established scientists just like any other scientist. Abū Maʿshar, in turn, gives us interesting insights into what distinguishes the experiences of the astrologer from those of laypeople and of the practitioners of other, lesser arts. Astrological experience has to adhere to scientific norms and

---

[35] Abū Maʿshar, *Great Introduction*, 6, 18, vol. 1, 662–663.

[36] Abū Maʿshar, *Great Introduction*, 6, 22, vol. 1, 672–673.

[37] Ibn al-Nadīm, *Fihrist*, trans. Dodge, 635.

[38] Other "testings" in Arabic are extant; several for medicine (e.g., *Kitāb Imtihān al-aṭibbāʾ* by Ḥunayn ibn Isḥāq, in Sezgin, *Geschichte*, 3:256) and at least one for astronomy, which al-Qabīṣī himself refers to in the preface to his *Imtihān al-munajjimīn*.

[39] A summary of the contents of the *Imtihān al-munajjimīn* is given in al-Qabīṣī, *Introduction to Astrology*, 5–7. A preliminary edition and English translation is given by Keiji Yamamoto: al-Qabīṣī, *Kitāb al-Imtihān*.

[40] Al-Qabīṣī refers to this work of his in the preface to his *Introduction to Astrology* (1.3, 19), but it has not been identified.

values, which are not accessible to everyone, have to be built over a long period of time, and must follow the rules of the craft. Thus the astrologer is in a privileged position to predict what will happen in the future.

2.4 Concluding Remarks

Abū Maʿshar goes out of his way to emphasize the nobility and universality of astrology compared to other practical arts, which are limited in their scope or are based entirely on observations of things in the sublunar world. The professionalism of astrology, or rather of practicing astrologers, distinguishes it from other crafts, which often lead to harm because of the inadequate knowledge of their practitioners.[41]

Abū Maʿshar refutes the arguments of several categories of laypeople who do not see the point of astrology, either because it is too difficult, or because its practitioners make errors, or because there is no way of avoiding the inevitable. These are some of the ten criticisms of astrology that Abū Maʿshar answers in *Great Introduction*, 1, 5, vol. 1, 106–149. But the main purport of his answers is that experiential knowledge, of which the astrologers have a wider range than any other practitioners, is tantamount to scientific knowledge. Indeed, when seeking to prove the legitimacy of astrology as an art and a science, Abū Maʿshar assigns a key role to experience. Experience of the world proves that astrology makes true claims about the dependence of terrestrial events on heavenly constellations. Such experiences are universal. They are available to everyone—but their meaning as signs is not. If other sciences and arts produce true knowledge based on reading signs, then astrology's predictions must be even more certain, as they work with unchanging, stable entities. Much scientific knowledge in astrology depends on experience, but not all scientific knowledge is experiential knowledge; analogy plays an equally important role.

Within the confines of his science, the astrologer's experience is unlike that of the common people. His experience is the product of much learning and training, as it is mediated by particular theories. Since the astrologer's task lies in reading signs in the world correctly so as to produce correct predictions for his client, the way in which he experiences the world must be considerably different from laypeople's. Astrological experience thus entails seeing the world in a very particular way. As a result, and as the variations in its name show, astrology can be regarded as both a *ṣināʿa* (craft or profession) and an *ʿilm* (science), and the good astrologer knows both the theory and the practice of astrology.

[41] [41] Abū Maʿshar, *Great Introduction*, 1, 2.31a, vol. I, 78–79.

Appendix

Capitulum in conceptione mulierum (from MS Paris, Bibliothèque nationale de France, lat. 16,204, p. 6). This capitulum is missing in Richard Lemay's edition of John of Seville's translation, Abū Maʿshar al-Balkhī, *Kitāb al-Mudkhal al-kabīr ilā ʿilm aḥkām al-nujūm*, vol. 5, 12, lines 372–391.

Plures quoque hominum accipiunt experimenta absque significatione stellarum ex rebus multis, ut mulieres obstetrices quoniam ipse sciunt utrum conceperit mulier vel non, et utrum conceperit masculum vel feminam. Sciunt etiam et ex partu \primo/ virginis utrum pariat mulier postea aut non et quot pariet. Et modicum errant in eo quod annuntiant ex hiis rebus per experimenta sua et per hoc quod audierunt ab antecessoribus suis de hiis que experte sunt in aliquo tempore. Scientia namque earum in muliere est utrum sit pregnans vel non, ut aspiciant mulierem quam suspicantur esse pregnantem, si viderint capita mammarum eius extensa et a colore solito mutata, scient ipsam esse pregnantem et ex signis pregnationis est ut aspiciant oculos mulieris, si fuerint concavi, fueritque in corpore eius laxatio et fuerit acuta visu, ac pupilla oculi fuerit clara, albedo quoque oculorum plena et spissa, scient ipsam esse pregnantem.

Many people also receive experiences without the indications of the stars; such as midwives, since they know whether a woman is pregnant or not, and whether she has conceived a male or a female. They know also from the first birth of a virgin whether the woman will give birth after this or not, and how many she will give birth to. And they err <only> a little in what they announce concerning these things, because of their own experiences and because of what they have heard from their predecessors concerning those things which they had experienced at one time. For the knowledge they have concerning a women is: whether she is pregnant or not; so they look at the woman they think is pregnant; if they see that the tips of her breasts have spread out or changed from their usual color they know that she is pregnant. A sign of pregnancy is that they observe the eyes of the woman: if they see that they have become sunken and if there is looseness in her body, and she has sharpness in her gaze, and the pupil of the eye is clear, also full whiteness and thick whiteness in her eyes, they know that she is pregnant.

Bibliography

Primary Sources

Abū Maʿshar al-Balkhī. 1995–1996. *Kitāb al-Mudkhal al-kabīr ilā ʿilm, Liber introductorii maioris ad scientiam judiciorum astrorum.* Arabic text and the two Latin translations, edited by Richard Lemay. 9 vols. Naples: Istituto universitario orientale.

Abū Maʿshar. 2019. *The Great Introduction to Astrology*, edited and translated by Keiji Yamamoto and Charles Burnett. 2 vols. Leiden: Brill.

Benjamin Dykes, ed. 2011. *The Book of the Nine Judges: Traditional Horary Astrology*, translated and edited by Benjamin Dykes. Minneapolis: Cazimi Press.

Hermann of Carinthia. 1982. *De essentiis*, ed. Charles Burnett. Leiden: Brill.

Ibn al-Nadīm. 1970. *Fihrist*. Translated by Bayard Dodge. New York: Columbia University Press.

Ptolemy. 1940. *Tetrabiblos*, edited and translated by F.E. Robbins. Cambridge, MA: Harvard University Press.

al-Qabīṣī (Alcabitius). 2004. *The Introduction to Astrology*, ed. Charles Burnett, Keiji Yamamoto, and Michio Yano. London: The Warburg Institute.

al-Qabīṣī (Alcabitius). 2005. *Kitāb al-Imtihān ("Book of Testing").* Preliminary edition and translation in Keiji Yamamoto, *Chuseini okeru Islam tenmongaku no Arabiago oyobi Latingo bunken no kenkyu*, 6–54. Kyoto: Kyoto Sangyo Daigaku.

Secondary Literature

Burnett, Charles. 2013. Doctors versus Astrologers: Medical and Astrological Prognosis Compared. In *Die mantischen Künste und die Epistemologie prognostischer Wissenschaften im Mittelalter*, ed. Alexander Fidora, 101–111. Cologne: Böhlau.

Burnett, Charles. Practical Magic: *nīranjāt* and Effective Recipes. Forthcoming in *Science and Craft: The Relations between the Theoretical and Practical Sides of the Occult and Esoteric Sciences in the Islamic World*, eds. Liana Saif and Godefroid de Callataÿ.

Sezgin, Fuat. 1970. *Geschichte des arabischen Schrifttums*, vol. 3. Leiden: Brill.

Vadet, Jean-Claude. 1963. Une défense de l'astrologie dans le *Madḫal* d'Abū Maʿšar al-Balḫī. *Annales Islamologiques* 5: 131–180.

Charles Burnett is Professor of the History of Arabic/Islamic Influences in Europe at the Warburg Institute, University of London. He works on the transmission of texts, techniques, and artifacts from the Arab world to the West, especially in the Middle Ages.

Chapter 3
Dream-Experience in Ibn Sīnā

Jules Janssens

Abstract Ibn Sīnā's conception of dreams is scattered over parts of different works, both philosophical and medical. In the former, he tries as much as possible to explain the experience of dreams on the basis of the functioning of the inner senses, especially the imaginary faculty, during sleep. This appears to be in line with Aristotle. Contrary to Aristotle, however, Ibn Sīnā affirms the possibility of supernal, divinely inspired influences—more particularly, influences related to the celestial souls, sometimes called "angels." Where Aristotle saw no real possibility of offering a causal explanation for prophecy in dreams, Ibn Sīnā is convinced he can offer one, by positing a congenerity between the celestial souls and our souls. On the medical side, however, he avoids any reference to experience based on divine intervention, limiting his comments instead to empirically observable facts. This complex approach to the experience of dreams can only be understood in the religious and scientific context of Ibn Sīnā's time.

Keywords Ibn Sīnā (Avicenna) · Aristotle · Galen · Arabic *Parva naturalia* · Dream experience · Divine inspiration

3.1 Introduction: The Historical Context

It is obvious that Ibn Sīnā—a most notable philosopher and physician in the Islamic Golden Age—worked out his views on dream-experience in the specific temporal and cultural context in which he was living, the fourth–fifth/tenth–eleventh centuries in the eastern part of the Islamic world. At that time, the reception of Greek thought especially, based on a vast translation movement of philosophical, scientific, and medical texts, had almost reached its end, and had already been integrated to a certain extent into the new Islamic (largely Arabic) context.

J. Janssens (✉)
De Wulf-Mansion Centre, KU Leuven, Leuven, Belgium
e-mail: jules.janssens@kuleuven.be

H. C. Erlwein and K. Krause, *Revisiting Premodern Islamic Science and Experience*, SpringerBriefs in History of Science and Technology,
https://doi.org/10.1007/978-3-031-76085-3_3

Strikingly, dreams played a non-negligible role in Islamic belief from its very beginning. In the Qur'an, explicit reference is made to three types of dreams or visions: symbolic, as attested in Yūsuf's dream-vision (*ru'yā*) of the prostration of eleven stars, the sun, and the moon to him (Q. 12:4); realistic, as illustrated by Muḥammad's dream-vision (*ru'yā*) of entering the Sacred Mosque (Q. 48:27); and pedagogically directional, as in Ibrāhīm's dream (*manām*) of sacrificing his son (Q. 37:102). These are all God-sent. However, in the first of the three cases, God will only reveal the proper interpretation (*taʿwīl*) to Yūsuf at a later moment (Q. 12:6). The second is presented as a real vision, conforming to the usual, objective way of seeing things in our daily lives. In the third case, the dream is explicitly described as a trial (Q. 37:106) and is thus pedagogical in nature, the expression of divine guidance. Although the first two kinds of dream-visions cannot be unconditionally identified with dreams, they undoubtedly have much in common with the third, since none involves a conscious move.

In addition to the references in the Qur'an, many dreams (or dream-visions) related to Muḥammad are mentioned in the biographical works on him (the *sīra* literature) and the *ḥadīth*.[1] Hence, quite naturally, a rich literature came to exist in which veridical dreams were presented as arising from divine inspiration, a source almost always mediated by angels.[2] This is confirmed by the ascription of confused dreams to devilish whisperings—that is, distorted dream-images caused by fallen angels. In sum, an important part of dream experience was valued by many serious Islamic scholars of the eighth–tenth centuries CE as having an ultimately divine origin, though most often with angelic (or devilish) mediation, and moreover as making accessible unknown things, including events in the future.

There is little doubt that the Arabic translation of *Oneirocritica* (The Interpretation of Dreams), written by the second-century CE professional diviner and interpreter of dreams Artemidorus, became a valuable contribution to dream interpretation, also in a specifically Islamic context.[3] It is less evident that philosophers—especially those claiming to be followers of Aristotle, including such critical ones as Ibn Sīnā—would accept a celestial origin for any kind of dream. Reading Aristotle's *On Dreams* and *On Divination* in their original Greek versions, one looks in vain for any reference to a possible celestial origin or influence. However, the case is radically different in the

---

[1] A serious survey of the most important testimonies in this respect is given by Fahd's seminal *La divination arabe*, 255–289. It is worth noticing that in such important collections of *ḥadīth*s as those of al-Bukhārī and Muslim, a specific book is dedicated to dreams (Books 91 and 42 respectively).

[2] Several dream manuals even mention the mediation of a specific "angel of dreams." See Albertini, "Dreams, Visions, and Nightmares," 171.

[3] Artémidore d'Éphèse, *Le livre des songes*, ed. Fahd, xi–xiv. Muslim readers undoubtedly found interesting Artemidorus's distinction between *oneiros* (Ar. *ru'yā*)—referring to the veridical dream as predicting the future and as including two fundamental kinds, in need or not in need of interpretation—and *enhypnion* (Ar. *aḍghāth*), referring to confused dreams that deal only with the present (note that the Arabic translation *aḍghāth* can easily be linked with the Qur'anic expression *aḍghāth aḥlām*; see below). Moreover, Artemidorus clearly links the veridical dream with a divine inspiration, rendered in the Arabic in terms of seeing angels of heaven (*malā'ikat al-samā'*) (Artémidore d'Éphèse, *Le livre des songes*, 29.3–29.4).

sole Arabic adaptation of the *Parva naturalia* that has survived.[4] There, the idea of the existence of God-sent dreams is expressed unambiguously, even if the universal Intellect, not God Himself, is designated as their immediate cause. In fact, God, in His providential act of creation, created the events that the dreams will foretell in a dual manner: once as universal "intellectual forms" in the intellect, and once as particular "perceptible forms" in the world. Yet it is more than puzzling how an intellect can contain particular forms.[5]

The Arabic adaptor seems simply to ignore this problem. Ibn Sīnā undoubtedly (given his acceptance of a Neoplatonically inspired emanation system) had no problem with the notion of a mediated divine communication of intellectual forms by a higher intellect, but in all likelihood he considered a communication of particular forms through an intellect to be utterly un-Aristotelian. As this chapter will show, he consequently tried to offer a reasonable solution to the problem by way of an influence of the celestial souls and their congenerity with our souls.[6] Whether or not he read Aristotle's *Parva naturalia*, Ibn Sīnā's work unmistakably valorizes much of Aristotle's scientific, that is, bio-psychological approach.

A very similar scientific approach also comes to the fore when Ibn Sīnā addresses dreams in a medical context. His explanation is set out exclusively in his era's scientific, empirical terms. With due prudence, this explanation can be labeled somatopsychic, or alternatively psychophysiological.[7] Its outspoken pathological and symptomatic bias clearly indicates its roots in Greek medicine, especially the Galenic tradition—though unlike his Greek predecessors, Ibn Sīnā avoids any reference in his medical writings to (God-sent) curative dreams.[8] However, it is worth stressing that in Galen's view, even inspired curative dreams do not transcend the (rational) science of medicine, for the knowledge of the gods is quantitively, but not qualitatively, different from that of the physician.[9]

---

[4] This adaptation has survived in acephalous form in a single Indian manuscript, Rampur 1752, fol. 7a–54b, where the second chapter (fol. 10r) gives as the title of the whole work "(Kitāb) al-ḥiss wa-l-maḥsūs." With regard to *On Divination*, see Hansberger, "How Aristotle," 50–64, but she insists that the combination of *On Dreams* and *On Divination* in a single part of the Arabic "adaptation" only vaguely represents the Greek text, "as far as it is justified to speak of 'representation' at all in this case" (ibid., 52).

[5] Hansberger, "Representation," 111.

[6] Ibn Bājja (d. 533/1138–39), with explicit reference to the end of the second part (which agrees with the division of the "adaptation") of *Kitāb al-ḥiss wa-l-maḥsūs*, ascribes veridical dreams to just a few exceptional people and describes them as divine inspirations or gifts. Ibn Bājja, *Tadbīr al-mutawaḥḥid*, 135–136. Even if his interpretation is not identical with Ibn Sīnā's, it shares the idea of a divine element and of the limitation of veridical dreams to an elite. When Ibn Bājja, without any hesitation, presents these ideas as proper to Aristotle (ibid., 136.10), it is obvious that he had no doubt about the authenticity of the attribution of the text to the Stagirite.

[7] The mention of "psychophysiological" as alternative qualification is inspired by Mazhar Shah, who, in his English translation of Book 1, translates the title of 1, 1, 6 as "faculties and functions," adding psychophysiology in parenthesis. Shah, *General Principles*, 125.

[8] See, e.g., Galen, *On Treatment by Venesection*, § 23, ed. Brain, 98.

[9] See Van Nuffelen, "Galen," 346.

3.2 Ibn Sīnā on Veridical Dreams of Supernal Origin and "Dream-Experience"

In the exposition on dreams in the *Kitāb al-nafs*, the "book on the soul" of his major philosophical *summa*, *al-Shifāʾ* (The Healing), Ibn Sīnā insists on the existence of veridical dreams that have their origin in the "Realm of the (Divine) Sovereignty" (*al-malakūt*) and that happen unexpectedly and suddenly.[10] He distinguishes between three major types of such veridical dreams, which I will now describe.

The first type are those needing no interpretation. Here, the imaginary faculty enables the soul,[11] when it is in "contact" with the "Realm of the Divine Sovereignty" during sleep, to firmly fix a given form (*ṣūra*), as it is, in the faculty of recollection (*dhikr*). This is strikingly similar to the way that the imaginary faculty's receptive perception of sensible forms in wakefulness prepares the rational soul to correctly fix that form in the faculty of recollection.[12] However, this enabling action of the imaginary faculty is possible only when the faculty is tranquil—purely receptive, not productive, and subjected to rational cogitation (*al-fikr al-nuṭqī*). Otherwise, the soul could never correctly establish what becomes apparent to it (*mā yalūḥ lahā*) on the basis of the imaginings of the imaginary faculty (*min takhayyulātihā*). Somewhat earlier in his explanation, Ibn Sīnā already expressed more or less the same idea, stating that in people who possess both a strong imagination (i.e., an imagination that is not overpowered by the senses and has full control over the form-bearing faculty) and a strong soul, the same state can occur in wakefulness as the one present in "the sleeper's perception [*idrāk*] of the things that belong to the reality of the Unseen [*mughayyabāt*] by ascertaining them [*bi-taḥaqquqihā*] either as they are in themselves or through [appropriate] images [*amthila*] for them."[13]

I will not deal here with the prophetic imaginary's capacity to perceive such things not only in the state of sleep, but also of wakefulness; instead, I will concentrate on the possibility that this specific perception can occur to any person in the state of sleep. On the terminological level, several terms deserve special attention: *mughayyabāt*, *taḥaqquq*, and *amthila*.

As Gutas has shown, the term *mughayyabāt* ("the Unseen") signifies, at least in this *De anima* context, all the particular events in the past, present, and future that

---

[10] [Ibn Sīnā], *Avicenna's De anima*, ed. Rahman, 175.13 and 178.6 respectively.

[11] Gutas, "Imagination," 345–347, rather convincingly shows that "soul" in this and similar cases refers not to the animal soul but the rational soul; here, see also Hasse, *Avicenna's De anima*, 150–160. However, as Gutas ("Imagination," 346–348), with due caution, observes, in one passage Ibn Sīnā's actual wording strongly indicates that the imagination itself is what effects the "contact" with the supernal world, when "the celestial bodies [...] drop forms into the imagination in accordance with [its] predisposition" (*Avicenna's De anima*, 180.5–6), although it could be claimed that this happens through mediation by the practical intellect aspect of the rational soul. This alternative interpretation is certainly plausible and worthy of attention, but in the absence of a direct textual basis, not absolutely evident.

[12] See *Avicenna's De anima*, 175.14–176.2.

[13] *Avicenna's De anima*, 173.14; translation in Gutas, "Imagination," 345 (slightly modified).

are included in the knowledge of the celestial souls.[14] However, Ibn Sīnā's focus lies on future events, as evidenced in his *Kitāb al-Hidāya* (Book of Guidance):

> Man, in the state of sleep, perceives [*yudriku*] the future particulars due to the turning away of his imagination and [all other of his] inner powers from sensation, and [in addition, due to] his "contact" [*ittiṣāl*] with the angels. The essences of the angels [present themselves] as if they were in an intelligible way the forms of being [or: existence], which encompass the ideas [*maʿānin*] of [all] events,[15] including [lit.: and of] the future.[16]

The term *taḥaqquq*, "ascertainment," in this context unmistakably implies the idea of learning something with a rationally founded certainty, which centers on the discovery of the middle term.[17] This latter qualification should, in my view, be understood in a broad sense, since mention is made of an act of perception that is not limited to things as they are, but also includes "images."

Amthila, "images," are clearly veridical ones that correctly and well reflect things as they are in themselves. These images correspond to the forms of being (or existence) that are present in the celestial souls, labeled "angels," in an intelligible way, as mentioned in the passage of the *Kitāb al-Hidāya* just quoted—where further confirmation can be found in the subsequent lines: "Because [man] imagines [*yataṣawwaru*] by means of these forms, they flow to his faculty of recollection [*dhikr*]. When they are firmly established and the imaginary faculty does not disturb them, they are recollected as they are."[18]

Given these terminological ramifications, one may wonder whether it is possible to speak of a dream-experience at all in this case and, if so, in what sense. Because the veridical dream (not in need of interpretation) involves a direct and unique perception of something certain, which is not accessible in this world on a purely sensible, empirical basis but only through the soul's "contact" with the supernal world, the technical sense of *tajriba*—"experience," systematic observation (and testing)[19]—must clearly be excluded. At best, this is a kind of experience in the less technical sense: personally encountering or undergoing something, without researching its

[14] See Gutas, "Empiricism," 33, 35; Gutas, "Imagination," 338–339, 345–347, 351–352; Gutas, "Intellect," 9–15. It is worth noting that in the metaphysical context, Ibn Sīnā ascribes this knowledge to God, while emphasizing that He is "the most knowing of the world of the Unseen [*al-ghayb*]." Avicenna, *The Metaphysics*, ed. Marmura, 290.17.

[15] I follow Lizzini's proposal to correct *al-kulliyāt* (universals) to *al-kāʾināt* (*avvenimenti*, events; Lizzini, "La Metafisica," 416 n. 227), but she does not seem to include the "future" in the notion of "events." I, however, am inclined to understand that all events, past, present and future, are intended—compare with Ibn Sīnā, *Avicenna's De anima*, 178.14–178.16: "The ideas of all the things that take place in this world—past, present and future—exist in the knowledge of the Creator and the angelic intellects in one respect, and in the angelic celestial souls in another respect." I therefore wonder whether the original text did not read *kull al-kāʾināt*? The explicit addition of "the future" would then constitute a particular emphasis on the fact that not only past and present events are involved.

[16] Ibn Sīnā, *Kitāb al-Hidāya*, ed. ʿAbduh, 295.1–4.

[17] Gutas, *Avicenna*, 214.

[18] Ibn Sīnā, *Kitāb al-Hidāya*, 295.4–5.

[19] For the understanding of *tajriba* as used here, see Gutas, "Empiricism," 46.

certainty by means of a syllogism (for like all veridical dreams, these ones arrive suddenly and unexpectedly). This might explain why Ibn Sīnā does not explicitly use the term *tajriba* when he presents dreams that do not need interpretation.

This stands in sharp contrast with the frequent use of the term, or derivates of the same root, in his account of the second type of veridical dreams. The veridical dreams of this second type can be labeled "interpretative" dreams, in which one dream offers the potential to interpret a previous dream:[20]

> The truest [*aṣaḥḥ*] dreams among them belong to men whose souls are habituated to veracity [*ṣidq*] and have subjugated the deceiving act of imagination [*al-takhayyul al-kādhib*]. The vast majority of those to whom it happens that one of their dreams is interpreted in another are those whose ultimate aspiration [*himma*] is preoccupied by what they have seen [*ra'ā*] [in their first dream]. Then, when they sleep, their preoccupation with it lasts according to the state in which it is [*bi-ḥālihi*], and, consequently, the imaginary faculty begins to imitate [*tuḥākī*] it [i.e., what they have seen] in the opposite way from how it imitated it the first time. [Example of Heracles.] When these warnings [*indhārāt*] were registered,[21] they happened according to how they had been interpreted for him [i.e., Heracles] in his [second, interpretative] dream. This has been experienced [*qad jurriba*] in others than him.[22]

Ibn Sīnā articulates the dream-activity related to the first dream, which is in need of interpretation, in terms of "seeing," not "perceiving." Furthermore, he points out that in the first dream, the imaginary faculty—by an act of imitation—creates an image from which the second dream—by an inverted act of imitation—returns to what was imitated in the first dream. This might give the impression that the imaginary faculty is sovereign in its action here, but that is not the case. On the contrary, at the beginning Ibn Sīnā stresses that this phenomenon can happen only in men of veracity who have (rational) control over it, hence whose imaginary faculty fully deserves to be described as the typically human cogitative faculty (*al-quwwa al-mufakkira*). Then and only then can the imaginary faculty, which is essentially an imitating faculty, not a sensing one, create through an act of imitation an image that corresponds well with what has been "seen" as a particular event in the celestial realm, simililarly to the reflection of something in a well-polished mirror.

Somewhat surprisingly, Ibn Sīnā concludes that others, just like King Heracles, have experienced a second dream that enables them to interpret a first one correctly. He uses the second form of the verb *jariba*, *jarraba*, of which *tajriba* is the verbal noun

[20] *Avicenna's De anima*, 176.18–177.15. In the first part (176.18–177.7), Ibn Sīnā presents such dreams in a framework where there sometimes seems to be no actual link with the supernal world. He explicitly defines these dreams as recollection and notes that they may even be due to the sole action of the imaginary faculty, cogitative, albeit qualified as "cogitative" (*mufakkira*); however, at the end of the part, he states that they cannot be precisely described. All this is puzzling and needs further examination.

[21] For the meaning of this term as referring to a celestial notification to humans of events that take or will take place on earth, see Gutas, "Intellect," 13.

[22] *Avicenna's De anima*, 177.7–15. Ibn Sīnā mentions the famous double dream of King Heracles in terms (partially) similar to the Arabic adaptation of the *Parva naturalia* (see ms. Rampur 1752, ff. 41r–42v; English translation in Hansberger, "How Aristotle," 63). Ibn Sīnā, *al-Mabda' wa-l-ma'ād*, ed. Nūrānī, 102.4, explicitly refers to the *Kitāb al-ḥiss wa-l-maḥsūs*, hence making it likely that he had access to it (or the original Greek–Arabic translation, now lost?).

(*maṣdar*). Here the question arises whether this use is based on a conscious choice, indicating a technical understanding. If so, what is the exact meaning involved?

Ibn Sīnā obviously wants to stress that having an interpretative dream is nothing unnatural, even if it happens only exceptionally. He underlines that this kind of dream happens to people who remain preoccupied by what they saw in a first dream, no oneirocritic having been able to offer a reasonable explanation. This rationally founded preoccupation, felt mainly in daytime, persists during sleep and causes a reactivation of the imaginary faculty, which is extremely active during sleep. More precisely, it pushes the imaginary faculty to turn back to the exact source of what formed the basis of its act of imitation and, when it has attained that source, to transmit it, as it is, to the faculty of recollection. The explanation offered is clearly located inside the framework of Ibn Sīnā's psychology of faculties, but it remains somewhat obscure why the imaginary faculty makes this inverse movement—inverse, that is, compared to the movement that ended in its earlier act of imitation.

Whatever Ibn Sīnā's exact argument here, it is beyond reasonable doubt that he intends his explanation to be a scientific one. As such, its object can be systematically observed. In this sense, one can attribute a technical sense to *jarraba* in this context: as systematic observation, albeit only in a very particular sense. Indeed, contrary to the usual understanding of *tajriba* in a this-worldly context, as exemplified by "scammony purges bile,"[23] systematic observation as proposed by Ibn Sīnā here does not concern the content. In the case of dreams, content is always unique, because it is based on the worries of a specific person about the interpretation of a specific dream. Instead, *jarraba* in this case is limited to structural resemblances on the formal side. Moreover, contrary to the usual notion of "experience" as implying the possibility of falsification related to specific circumstances, such as climatic conditions with regard to the efficacy of certain drugs, the interpretative dream offers a truth that is always absolutely certain. This is understandable, given that the interpretative dream deals with particular facts and has its origin in a celestial source—a celestial soul. Finally, the interpretative dream is not based on a fully conscious and rational analysis of rational data. Certainly, a rational element is involved, insofar as the dream only comes to people who fully master their imaginary faculty and keep it under rational control. Due to the congenerity of the human soul with the celestial souls, human beings' imaginary-cogitative faculty is able to understand some particulars as implied in the intelligible universal order when, in their dreams, they have "contact" with the celestial souls, which have an encompassing knowledge of all particulars in that order. From this perspective, the interpretative dream can be labeled a special case of a dream that is not in need of interpretation, and as such it entails a very weak

---

[23] In the *Kitāb al-Burhān* (Book of Demonstration) of his *Shifā'*, Ibn Sīnā uses this medical example of a laxative to illustrate the degree of certainty of knowledge that can be derived from "experience"—not absolute certainty, but only in a qualified, context-dependent way. See Janssens, "'Experience,'" 54–57.

sense of experience. Therefore, Ibn Sīnā's use of the term *jarraba* should probably not be overevaluated in this case.[24]

The third type of veridical dreams are those in need of an interpretation, namely whenever the imaginary faculty replaces the original object of the dream with an imitation (*muḥāka*)—hence, when the imaginary faculty is dominant in the transmission of the dream content.[25] These dreams are in principle open to everyone; in fact, "there is not a single person who has not had his share of [veridical] dreams [*ru'yā*] [while asleep]."[26] Such dreams can in principle lead to the discovery of the truth with regard to all things past, present, and future, but conversely they can, in the worst case, end up as what the Qur'an labels "muddled dreams" (*aḍghāth aḥlām*): when the action of the imaginary faculty "overpowers the soul in [such] a way that it distracts the soul from completing what it has seen" (i.e., in the "Realm of the [Divine] Sovereignty").[27]

In his *Kitāb al-Hidāya*, Ibn Sīnā briefly but significantly describes the process that underlies the generation of such a dream, as well as how one can arrive at a correct interpretation of it:

> But if the imaginary [faculty], as is its habit, starts to imitate them [i.e., the forms seen due to a "contact" with the "angels"] in the same way as it imitates by means of similar or contrary [things]—and such is its nature—what is present to it [...], then—[namely] when it imitates what it [originally] saw [*ra'at*]—this [imitation] will sometimes be firmly embedded in the form-bearing [faculty] and the original [form] will disappear. And so there will be a need for an interpretation in which one returns through an act of imagination [*takhayyul*] [to the original form] by establishing a match between parts of the imitations and parts of what was seen [*ru'iyā*] first.[28]

As already became evident in my presentation of the interpretative dream, the imaginary faculty's imitation of forms "seen" by it in the celestial souls implies a risk of obscuring the profound truth involved in the original "vision." In fact, as Ibn Sīnā now stresses, this imitation does not automatically involve similarities with those original forms but can proceed by presenting opposite images. Since such imitation can involve several acts—as implied by the use of the plural form ("imitations") at the end of the fragment—this immediately explains the possibility of "muddled dreams." Also important is that the imitated form is located in the form-bearing faculty, that is, the lower faculty of memory. That faculty preserves first and foremost the form abstracted from the sensible data as united in the common sense.

[24] As far as I can see, the only other work where the phenomenon of interpretative dream is explictly referred to is the natural part of *al-Mashriqiyyūn*, although limited to the very first lines of the exposition in *Kitāb al-nafs*, 4, 2; see Özcan, "İbn Sînâ'nın el-Hikmetu'l-Meşrikiyye," 170.15–17. The example of Heracles has—most likely, consciously—been omitted, including the last affirmation about others' experiencing it. Hence, all in all, it receives little attention.

[25] *Avicenna's De anima*, 176.4–176.10, 15–17.

[26] *Avicenna's De anima*, 174.1; English translation Gutas, "Imagination," 349.

[27] *Avicenna's De anima*, 176.11–176.15; for the expression in the Qur'an, see Q. 12:44 and 21:5.

[28] Ibn Sīnā, *Kitāb al-Hidāya*, 295.6–296.1.

However, the imitated form now takes the place of the abstracted form, and thus acquires a "quasi-sensitive" character.[29]

Finally, just as for the interpretative dream, discovery of the true sense of the dream in need of interpretation is articulated in terms of a process of returning from the imitation to the originally "seen" form. The proposed procedure is that of establishing a comparison between a given imitation and a given form as seen in the dream.[30] Unfortunately, Ibn Sīnā's exposition does not indicate how such a "match" can be made in practice—not least because he seems to imply that in such cases, there is still an awareness of what was originally seen in the dream, even after the originally "seen" form has been imitated. This awareness seems to indicate an acceptance of fixed dream-symbols, which could enable comparative study of different dreams and their ultimately correct interpretation by serious oneirocritics, thus offering a basis for the proposed matching—but Ibn Sīnā remains silent on this point.

A crucial new element regarding how to interpret a veridical dream once the original form perceived in a given dream has been obscured comes to the fore in the *Dāneshname-ye ʿAlāʾī* (Book of Science [or Wisdom/Philosophy] for ʿAlāʾ [al-Dawla]), Ibn Sīnā's only Persian *summa.* This time, too, Ibn Sīnā stresses right away that in cases where the original content of the dream is far from clear, it is nevertheless possible that "by a trick of cogitation [*ḥīlat-e fikr*] you will restore [your] earlier thought."[31] But, again as in his previous work, Ibn Sīnā expresses himself in a very enigmatic way, especially when he presents the instrument for achieving the return to the original vision as a "trick of cogitation."

As I have already mentioned, cogitation undoubtedly refers to the inner sense of imagination as specifically related to man, as a faculty explicitly at the service of the intellectual faculty. Hence, Ibn Sīnā clearly implies a full reorientation toward reason here. But what kind of trick can make an imaginary faculty, completely devoid of any control over reason in its action, conducive to accepting a new submission to reason? Is it by making man aware that unbridled imagination is de facto destructive for his life, both in this world and in the hereafter, by facilitating irrational and consequently immoral acts? Whatever the trick actually is, at the end of the chapter Ibn Sīnā underlines that the interpretation of a veridical dream is based on conjecture (*takhmīn*) and is established by "experience" (*tajribat*):

> The core of this interpretation [*taʿbīr*] [i.e., of a veridical dream where imagination has obscured the original form] is that you say: "What may I have seen in the world of the Unseen in such a manner that the faculty of imagination passed from it to something else?," for example, "What did I see in such a manner that the faculty of imagination made it into a tree?" Hence, [such] interpretation is for the most part by conjecture and it is performed by experiences [*tajribathā*]. To each [specific] nature belongs another habit, and, according to

---

29 Compare *Avicenna's De anima*, 170.1–170.14, where it is affirmed that such form can, in the case of little or no intellectual control, even be impressed in the common sense.

30 With due prudence, I interpret "parts" in the sense of "a given," because I feel unable to make any serious sense of the affirmation if it is understood in its usual sense.

31 See Ibn Sīnā, *Dāneshname-ye ʿAlāʾī*, *Ṭabīʿāt*, ed. Meshkāt, 133.12–134.4.

> each time [or: season? *faṣlī*] and [each] state, a different imitation [*muḥāka*] belongs to the faculty of imagination [*mutakhayyila*].[32]

In affirming that the interpretation of dreams is most often based on conjecture, Ibn Sīnā stresses that such interpretations can seldom be seen as expressing certainty, not least because they must deal with context-dependent imitations, made by the faculty of imagination, that are based on habits, times, or states. For him, dream-interpretation almost always includes a degree of conjecture—at least minimal, but otherwise unspecified, and hence perhaps even relatively high. This is because there are no universally valid correspondences between the original seen forms and the remembered dream-images, which result from imitations made by the imaginary faculty.

Contrary to what appeared to be a plausible hypothesis in the framework of the *Kitāb al-Hidāya*, the acceptance of fixed dream-symbols is evidently excluded in the *Dāneshname-ye ʿAlāʾī*. Ibn Sīnā does not rule out the possibility of discovering the originally transmitted truth, namely by experiences. But, again, the question arises of what this term "experience" precisely means in our context. Clearly, it entails a strongly subject-related dimension: the dreamer's proper state, both psychological (i.e., the extent to which the dreamer lacks rational control over the faculty of imagination, as evidenced by the arrival of imitations) and physical (i.e., the habits, times, and states in which the dreamer lives). As far as the physical aspect is concerned, one easily detects a close similarity with the scammony case, thus context-dependency. However, an additional element missing in the discussion of the experience of drugs is the relative power of the inner sense of imagination—the greater that power is, the more difficult it becomes to find the original content of what was revealed in the dream-vision before it was subjected to imaginary imitations.

The "I" dimension occupies a central place in both the need for and the activity of experience: "What may *I* have seen?" The use of the plural "experiences" is most likely due to the fact that in a well-defined framework of time and place, a given person can find out the true meaning involved in a dream that came from a supernal source, probably by rationally comparing a particular dream with earlier dreams and concentrating on how their truth became manifest in their own particular contexts, including—most importantly—the dreamer's own mental disposition. One might call this a kind of systematic, comparative observation, something that also lies at the basis of "experience" in relation to scammony and other drugs, but whose goal is not to demonstrate what has been observed but to discover facts that are not accessible to ordinary, sensible human observation. It is obvious that in Ibn Sīnā's view, even without any further clarification as to how one can return to the original observation, a rational explanation of a veridical dream is always possible. Unfortunately, however, we find no detailed explanation of how an original form can be detected on the basis of a given imitation. The *Dāneshname-ye ʿAlāʾī* offers no real solution to this difficulty, any more than did the *Kitāb al-Hidāya*.

---

[32] Ibn Sīnā, *Dāneshname-ye ʿAlāʾī*, *Ṭabīʿāt*, 134.6–135.1. Note that the idea of conjecturing is already present in a similar context in his early work, *al-Mabdaʾ wa al-maʿād*, ed. Nūrānī, 118.23, but without being accompanied by that of "experience."

However, in the *Ishārāt wa-l-tanbīhāt* (Pointers and Reminders), *namaṭ* 10, there are further elements of clarification regarding veridical dreams requiring or not requiring interpretation. In Chapter 8, Ibn Sīnā says:

> "Experience" [*tajriba*] and "[syllogistic] reasoning" [*qiyās*] compound consentaneously [*muṭābiqūn ʿalā*][33] to [establish] that in the state of sleep, the human soul possesses [the capacity] to acquire something [lit.: an acquisition] from the world of the Unseen [*al-ghayb*]. [...] As to "experience," hearing [from others] [*tasāmuʿ*] and [personal] acquaintance [*taʿāruf*] certify it. There is no single human being who has not experienced [*jarrabat*] this by himself [in the form of] experiences [*tajārib*] that inspired [*alhamat*] him with assertion [*taṣdīq*], excepting for one who has a corrupt temperament and weak powers of [exercising] imagination [*takhayyul*] and recollection [*tadhakkur*].[34]

In the opening lines of this passage, Ibn Sīnā affirms that "experience," alongside syllogistic or analogical reasoning,[35] proves it is possible for the human soul to acquire knowledge of something (i.e., particular future events) from the supernal world.[36] Here "experience" could start to achieve a methodological significance, namely as a means to possess absolute certainty regarding the possibility for humans to receive information about otherwise unknown—or even, on a strictly earthly basis, unknowable—facts that comes directly from the supernal world. The explicit reference to an interpersonal element ("hearing from others") cannot have any validity if there is no way to check the veracity of their testimonies.

Even if the modalities of such verification are not specified, Ibn Sīnā clearly suggests that it can be carried out by comparison with one's own "experience" of the reception of information that comes directly from the supernal realm. Yet in such a subjective account, "experience" risks losing its methodological character. Ibn Sīnā seems to have been aware of this risk and, in his final affirmation, therefore uses the plural form "experiences." He thus implicitly refers to a reception—in principle utterly subjective, but observed many times both by oneself and by others—of a specific kind of knowledge that is accessible to humans only through contact with the world of the Unseen. Hence, in spite of the impossibility of comparing these dreams systematically, because they concern particular events and not universals, Ibn Sīnā does not hesitate to qualify the overall process in terms of "experience,"

---

33 For this understanding of *ṭābiqa ʿalā*, see Lane, *Arabic-English Lexicon*, 5:1825.

34 Ibn Sīnā, *Kitāb al-ishārāt wa-l-tanbīhāt*, ed. Forget, 209.16–210.3.

35 This (syllogistic/analogical) reasoning may consist in the search for middle terms, as usually is the case, but Ibn Sīnā (*Kitāb al-ishārāt wa-l-tanbīhāt*, ed. Forget, 215.13) adds: "or what resembles them," detailed by al-Ṭūsī (see al-Ṭūsī, *Sharḥ al-Ishārāt*, ed. al-Āmolī, 1136) as: "the conditional part in the conditional syllogisms," or what resembles the "middle (term)" (*al-awsaṭ*) in inductions and "arguments by example/paradigmas" (*tamthīlāt*). Unfortunately, Ibn Sīnā does not explain how he understands *qiyās* here, but implies that one can find further elucidation of them in the notes (*tanbīhāt*) that follow. However, since I concentrate in this paper on the issue of "experience," I will not analyze this in depth here.

36 This explication is based on al-Ṭūsī, *Sharḥ al-Ishārāt*, 1110; the same text is present in Ibn Kammūna's commentary, where, however, it it added that when there are multiple attestations to the "hearing of others," which confirm such a knowledge, and this either on the basis of veridical dreams that need no interpretation or ones that are open to explanation (*tafsīr*, Qur'anic exegesis), any form of deceit must be excluded (Ibn Kammūna, *Sharḥ*, ed. Malekī and Noorisefat, 473).

understood in a methodological sense, namely, as based on repeated occurrences of a reception of the higher world.[37]

Ibn Sīnā eventually mentions that such experiences "inspire assertion." This expression occurs, as far as I can see, nowhere else in Ibn Sīnā's works. The word *taṣdīq*, "assertion," is part of Ibn Sīnā's technical terminology: it expresses a judgment concerning the truth-value of a proposition.[38] Here in the *Ishārāt*, however, experience is not said to be the unique and direct source of assertion; it is only said to "inspire" assertion. The verb *alhama* and related terms such as *ilhām*, "inspiration," are used by Ibn Sīnā, though rather seldom and mostly (not exclusively) in relation to a higher, supernal source.[39] That "inspiration" in our context implies a celestial origin is evidenced in the following affirmation in Chap. 22:

> What is retained with stability of the kind of trace that is under discussion,[40] in the faculty of recollection [*dhikr*], whether in the state of wakefulness or of sleep, does not require exegetical explanation [*ta'wīl*] or interpretation [*ta'bīr*], [regardless of whether it is] an inspiration [*ilhām*], a clear revelation [*wāḥy ṣurāḥ*], or a [clear] dream [*ḥulm*]. But what [of such a trace] has ceased, while successive imitations of it [*muḥākiyyatihi wa tawāli'*] remain, requires one of those two [i.e., exegetical explanation or interpretation], while this varies depending on individuals, times, and habits: revelation requires [exegetical] explanation and dream [*ḥulm*] interpretation.[41]

Ibn Sīnā particularly emphasizes that "inspiration" implies the correct preservation of what he has previously labeled "a spiritual trace," and this specifically in the faculty of recollection. As such, this kind of trace can be identified with an engraving (*naqsh*) in the soul of something that comes from the realm of the Unseen.[42] To a

---

[37] Unlike in the *Burhān* of the *Shifā'*, one looks here in vain for the presence, in the idea of "experience," of a hidden syllogism or a causal relation that is valid for a retricted field of application. See Janssens, "'Experience,'" 56 and 61.

[38] See Avicenna, *The Healing, Logic*: *Isagoge*, ed. Di Vincenzo, 278. In his *al-Najāt* (The Salvation) (ed. Dānishpazhūh, 7.4–5), Ibn Sīnā specifies that "[genuine] assertion is acquired by means of [syllogistic] reasoning [*qiyās*], or what is similar to it.".

[39] These instances cover very different issues: the celestial conception of the good as a source for transformations of natural matters (Avicenna, *The Metaphysics*, 10, 1, 361.12–14 Ar.); the acquisition of knowledge of self-evident intelligibles due to divine inspiration (Landauer, "Die Psychologie," 361.9–12) or of *ḥads* due to "contact" with the Agent Intellect (Ibn Sīnā, *al-Najāt*, 341.1–6—this affirmation is absent in the otherwise mainly corresponding texts of *Aḥwāl al-nafs*, on the one hand, and *Kitāb al-nafs* of *al-Shifā'*, on the other); the direct grasping of *ma'ānin* by the faculty of estimation due to a divine inspiration (*Avicenna's De anima*, 183.16–18, 184.2, 205.6—in animals also called [184.5] "instinctive inspirations" [*ilhāmāt ghariziyyāt*] and in men [206.1] "a kind [*ḍarb*] of inspiration," related to human beings' ability or inability to foresee their future in terms of required actions); a non-conventional significance of expressions based on a divine inspiration (Ibn Sīnā, *al-Shifā'*, *al-'Ibāra*, ed. Khuḍayrī, 2.15, 3.6); an inspiration that resembles revelation (explictly said to be due to an action of the spiritual substances), accorded to pious and just people (Ibn Sīnā, *Aqsām*, ed. Shamsaddīn, 269.17–18). In this list, there is no direct link with our topic of dreams (or, relatedly, sleep), but, as I will show, the *Ishārāt* fills this gap.

[40] The reference is to the "spiritual trace," *al-athar al-rūḥānī*, discussed before in Chapters 19 and 21.

[41] Ibn Sīnā, *Kitāb al-ishārāt wa-l-tanbīhāt*, 216.15–217.3.

[42] Ibn Sīnā, *Kitāb al-ishārāt wa-l-tanbīhāt*, 10, 18, 214.4–6. The notion *naqsh*, "engraving" (or related terms of the same root *n-q-sh*) is attested in several other chapters as well, e.g., 9–10, 12–14,

certain extent, though not completely, inspiration is similar to the case of a dream that is not in need of interpretation and the case of a clear revelation, insofar as it implies a full grasp of what becomes available to our human souls from the world of the Unseen. Inspiration absolutely requires the imaginary faculty to remain fully under the control of the rational soul. In view of this, the affirmation of "inspiring assertion" at the end of 10, 8 seems to concern primarily people who have not only a powerful (rational) soul, but also powerful imaginary faculties of imagination and recollection—a group mainly composed of prophets and scholars. Only in their case does the congenerity between the human soul and the celestial souls appear almost complete. If *ḥads*, the correct guessing of the middle term, permits this small elite to discover the intellectual truth immediately, or almost immediately, the experience of the "clear dream," which does not require interpretation, allows the elite to fully grasp immediately, or almost immediately, the exact content implied in its vision.

Note that this in no way implies any form of mysticism, any more than the acceptance of *ḥads* does. But unlike the case of *ḥads*, the clear dream does not deal with universals, since these can only be arrived at through an illumination by the Agent Intellect. Rather, the clear dream is about particular events, knowledge of which is transmitted to the imaginary faculty from the celestial souls. The imaginary faculty transforms this knowledge into images that it deposits either in the faculty of recollection, in the form of the original images including their *maʿānin*, or in the common sense, in the form of imitations of the original images in the guise of sensible forms (which are here based not on sense-perception but on "inner observation," *mushāhada*).[43]

In neither of these cases can clear dreams be derived from purely intelligible matters. This is said expressly in Chap. 24 of the *Ishārāt*:

> Know that the way to profess and attest to these things [namely, related to the world of the Unseen] is not [to say], "they are merely plausible conjectures [*ẓannūn*] at which one arrives from intelligible matters only," even though this would have been something to rely upon had it been [the case]; rather they are experiences [*tajārib*] of which, once confirmed, one seeks the causes.[44]

---

and 19. Regarding the vagueness of this notion as a metaphor, as well as the various ways in which it is manifested according to Ibn Sīnā, see Gutas, "Imagination," 351–53. Bennet, "Avicenna's Dreaming," 94, interestingly notes that through this notion (translated by Bennet as "inscription") Ibn Sīnā establishes an "epistemological continuum" in which—according to Fakhr al-Dīn al-Rāzī's interpretation as expressed in his commentary, as Bennet mentions—"an instance of knowledge may be realised in different states: that is, as a universal and as a particular." Regrettably, Bennet pays no attention to the role of the celestial bodies (although this is clearly referred to in the *Ishārāt*, 10, 9).

[43] See Ibn Sīnā, *Kitāb al-ishārāt wa-l-tanbīhāt*, 10, 14, 212.19, and 10, 15, 213.9–213.10. Gutas, "Empiricism," 45–46, translates *mushāhada* as "Experience," noting that for Ibn Sīnā it implies observation with all the senses and the mind. However, since in our context only the imaginary faculty is in question, I find "inner observation" is justified.

[44] Ibn Sīnā, *Kitāb al-ishārāt wa-l-tanbīhāt*, 218.9–218.12; English translation in Gutas, "Imagination," 352.

Not conjectures but experiences form the basis for attesting the reality of things of the Unseen. Since these dreams are not uniquely based on intelligible matters,[45] they are in need of interpretation and hence relate to veridical dreams, as evidenced by the qualification "once confirmed." This qualification, moreover, is clearly not based on any intellectual activity, but only forms a starting point for such activity—specifically, the search for a causal explanation of why and how the dreams occur. This is underlined in the next passage of the same text:

> One of the instances of happiness that is vouchsafed to lovers of profound reflection [*istibṣār*][46] is both that these states [related to such experiences] should occur in them personally and that they should observe them [*yushāhidūhā*] several times successively in others, to the point that all this becomes an "experience" [*tajriba*] of relevance to the establishment [*ithbāt*] of something marvelous that has sound existence [*kawn wa-ṣiḥḥa*], and an incentive to seek its cause.[47]

This way of proceeding in the acquisition of knowledge is rather different from the one used for determining the efficacy of drugs. In the case of drugs, an element—though limited—of controlled testing is present,[48] but it is absolutely missing in the experiencing of dreams, because they happen unexpectedly and unconsciously. However, a comparative observation between several of one's own dream-experiences, and afterwards between these and the dream-experiences of others, enables the multiplicity of unique, personal experiences (all related to particular events) to be reduced to what is obviously common to all of them. Ibn Sīnā explicitly and unambiguously labels this process of comparison *tajriba*, "experience."

Thus, experience is explicitly endowed with a defined methodological significance: as repeated observation. Note, however, that repeated observation does not in and of itself lead to direct knowledge of the Unseen. In fact, it only proves that the Unseen exists, since human beings—in principle, all of them—sometimes have access to some elements of the Unseen, as indeed is attested by the observable regular occurrence of veridical dreams across time and place. Moreover, the search for the cause of our access to elements of the Unseen is incontestably intellectual in nature, since any causal explanation is inconceivable without actual supervision by the intellect, as Ibn Sīnā himself emphasizes: "When it [this marvelous something] is elucidated, the [rational] soul finds reassurance [*iṭma'anna*] in the existence of

---

[45] In this sense, they most likely have their origin in the celestial souls, which "have a mode of doing things with particular meanings by way of an apprehension which is not purely intellectual, and that is for the like of [such celestial souls] to arrive at an apprehension of particular events" (Avicenna, *The Metaphysics*, 359.34–360.2).

[46] Gutas adds the specification "philosophical." It almost certainly corresponds to what Ibn Sīnā de facto had in mind, but given the allusive nature of the *Ishārāt*, I prefer the more neutral "profound." Note that the choice of the term *istibṣār* may have been influenced by Q. 29:38, where, at the end of the verse, the word *mustabṣirīn* appears, rendered by Yusuf Ali, *The Holy Quran*, 1038, as "gifted with Intelligence and Skill."

[47] Ibn Sīnā, *Kitāb al-ishārāt wa-l-tanbīhāt*, 218.2–218.5; English translation in Gutas, "Imagination," 352–353.

[48] Janssens, "Ibn Sīnā on 'Experience,'" 309.

these causes, while the estimation [*wahm*] yields and does not oppose the intellect in its vanguard function relating to the [investigation of] these [causes]."[49]

3.3 Dreams as a Diagnostic Tool

In the *Kitāb al-nafs* of the *Shifāʾ*, Ibn Sīnā indicates that there is a close link between the dreamer's temperament and the truth-value of his dreams: the more equilibrated one's temperament, the truer are one's dreams.[50] Hence, in cases where the imaginary faculty imitates natural things that are near to it—and with regard to which Ibn Sīnā previously stated that, "by the act of blending the powers of the humors," they "belong to the pneuma [*rūḥ*] that uses the form-bearing faculty and the imaginary [faculty] as its vehicle"[51]—the dream is conditioned by the humoral (that is, physiological) nature of the dreamer's temperament.

The intimate relation between a given type of dream and the different humors is discussed in greater detail in the section on signs and symptoms in *al-Qānūn fī l-ṭibb* (The Canon of Medicine), Book 1, part 2, lesson 3, especially Chapters 3–4 and 6–7.[52] In this section, Ibn Sīnā enumerates ten kinds of signs (*ʿalāmāt*) that reveal different temperaments. Discussing the last kind, signs derived from the actions and passions of the psychic faculties, he makes an explicit reference to dreams:

> Dreams [*al-aḥlām wa-l-manāmāt*] also belong to this [last] kind [of signs]: a person in whom the hot temperament [*mizāj*] dominates sees [himself in his dream] as warming himself by means of the fire [*bi-nīrān*] or as sunning himself [reading: *bi-tamashshus*], and one in whom a cold temperament dominates sees [himself in his dream] as if he is cooled with snow or submerged in cold water. The possessor of each humor [*khilṭ*] sees [in his dream] what is congenerous with the humor [prevailing in] him, as is said.[53]

Here Ibn Sīnā, as he did in the *Kitāb al-nafs* of the *Shifāʾ*, regards given types of dreams as symptomatic of the dreamer's temperament. However, where the *Kitāb al-nafs* makes a direct connection between the truth-value of a dream and the temperamental state of the dreamer, the *Qānūn* insists on a connection of physical similarity between the nature of a prevailing temperament and the type of dream-image.[54] This is confirmed in the fourth chapter, where pleasant and amusing dreams are said to

[49] Ibn Sīnā, *Kitāb al-ishārāt wa-l-tanbīhāt*, 218.15–17; English translation in Gutas, "Imagination," 353.

[50] *Avicenna's De anima*, 180.18–181.1.

[51] *Avicenna's De anima*, 179.9–10.

[52] The fifth chapter discusses the exceptional cases of abnormal temperament which, in Ibn Sīnā's view, are manifested by serious bodily malformations. Just as he ascribes to different types of dreams a diagnostic value with regard to determining a state of imbalance in the temperaments, Ibn Sīnā ascribes to these malformations the very same kind of diagnostic value.

[53] Ibn Sīnā, *al-Qānūn fī l-ṭibb*, 1, 2, 3, 3, ed. Habibi, 428.1–428.3.

[54] Hence, the type of link that characterizes the connection between dream-image and temperament is expressed in physical terms in the *Qānūn* and, quite differently, in terms of truth-value in the *Kitāb al-nafs* of the *Shifāʾ*.

indicate a balanced temperament, and in the sixth, where a given type of dream is linked with the presence of a quantitative (related to the vessels) or qualitative (related to vitality) plethora. Its strongest confirmation is in the seventh chapter. There, a given type of dream is identified as symptomizing the dominion of one of the four humors over the three others: dreams of red or bloody, white or watery, yellow or fiery, dark or frightening things are related to the predominance of blood, phlegm, yellow bile, and black bile (*sawdāʾ*) respectively.[55]

In line with the Galenic tradition,[56] Ibn Sīnā clearly accepts that each specific dream experience reveals the current state of temperament of the dreamer. Inside the specific framework of medicine, the dream experience has an unequivocally articulated scientific basis, namely the physiology of the four humors—the somato-psychic or psychophysiological doctrine that was commonly accepted in Ibn Sīnā's time.[57] As far as I can see, the only other place in the *Qānūn* where Ibn Sīnā speaks of dreams is in the section on the diseases of children and their treatment. There he posits an over-full stomach (due to excessive voracity) as a cause of nightmares in small children, hence attributing a diagnostic value to dreams once again.[58] Even if Ibn Sīnā never explicitly uses the term "experience" in his presentation of the dream as a diagnostic tool, it is obvious that he could have labeled it as such in the sense of systematic observation of these types of dreams as signs (in this case, diagnostic signs).

3.4 Conclusion

It is an open question whether Ibn Sīnā's proposed explanation of what he himself labels "dream-experience" is satisfactory in all respects. But there is no possible doubt that he did regard this as a very special kind of experience: one that is due to influences coming from the "Realm of the (Divine) Sovereignty." Given the congenerity of the human soul with the celestial souls, who know the particulars as included in the providential order of the universe in a particular way, it seems that the human soul is sometimes able to receive, and even to grasp correctly, elements of this encompassing knowledge. However, the imaginary faculty also plays a crucial role in the reception of the knowledge of such particular events, because it is the most active faculty of the soul during sleep.

Unfortunately, Ibn Sīnā says little about the precise modalities of such reception. He is, though, crystal clear that influences from the Realm of Sovereignty are the

[55] Ibn Sīnā, *al-Qānūn fī l-ṭibb*, 1, 2, 3, 3, 434.2–3, 9–10; 435.1–2, 8–9; these wordings have much in common with a fragment, attributed to Galen, that is extant only in Arabic. See Gätje, *Studien*, 139.

[56] It is beyond reasonable doubt that Galen influenced Ibn Sīnā, but further research is needed in order to fix more precisely the extent of that influence, and whether it was based directly on Galen's writings or through intermediary sources.

[57] See note 7 above.

[58] Ibn Sīnā, *al-Qānūn fī l-ṭibb*, 1, 2, 3, 3, 561.4–561.6.

only possible way to arrive at a correct knowledge of future particulars.[59] Dream experiences are presented as subjective experiences that can in principle happen to anyone at all, although Ibn Sīnā adds a strict condition: reception is only possible in the absence of a faculty of imagination that entirely irrationally multiplies imitations and of an imbalanced temperament. If this condition is fulfilled, comparative study can reveal a common pattern in all these experiences, opening the door to genuinely intellectual research on the precise cause of the deeper, underlying reality. Here, Ibn Sīnā is in search of a scientific explanation. That he took this quest seriously becomes manifest in the way he deals with dreams in his medical works, where he gives his explanation exclusively in his era's scientific, empirical terms: humoral balance or imbalance, and an overloaded stomach in children. In sum, Ibn Sīnā does not reduce the dream-experience to an inexplicable phenomenon but gives it a scientific explanation, based on the faculty-oriented conception of the soul in his philosophical psychology and his conceptions of humors and digestion in medicine.

Bibliography

Primary Sources

Artémidore d'Éphèse. 1964. *Le livre des songes, traduit du grec en arabe par Ḥunayn b. Isḥāq*, ed. Toufic Fahd. Damas: Éditions d'Amérique et d'Orient.

Galen. 2010. On Treatment by Venesection. In *Galen on Bloodletting: A Study of the Origins, Development and Validity of His Opinions, with a Translation of the Three Texts*, ed. Peter Brain, 67–99. Cambridge: Cambridge University Press.

Ibn Bājja (Avempace). 2010. Tadbir al-mutawaḥḥid. In *La conduite de l'isolé et deux autres épîtres*, ed. Charles Genequand, 121–182. Paris: Vrin.

Ibn Kammūna. 2021. *Sharḥ al-ʿuṣūl wa-l-jumal fī muhimmat al-ʿilm wa-l-ʿamal (Sharḥ al-Ishārāt wa-l-tanbīhāt li-Ibn Sīnā), taʿliq Sharaf al-Dīn l-Masʿūdī wa- Alī Ibadī Shahrūdī*, ed. Muḥammad Malekī (Jalāl al-Dīn) and Marzieh Noorisefat (Noora). Tehran: Mirase-Maktoob.

Ibn Sīnā (Avicenna). 2021. *"The Healing, Logic: Isagoge": A New Edition, English Translation and Commentary of the Kitāb al-Madḫal of Avicenna's Kitāb al-Šifāʾ*, ed. Silvia Di Vincenzo. Berlin: De Gruyter.

Ibn Sīnā (Avicenna). 2005. *The Metaphysics of the Healing*. A parallel English-Arabic text, translated, introduced and annotated by Michael Marmura. Provo, UT: Brigham Young University Press.

Ibn Sīnā. 1892. *Le livre des théorèmes et des avertissements / Kitāb al-ishārāt wa-l-tanbīhāt*, ed. Jacques Forget. Leiden: Brill.

Ibn Sīnā. 1951. *Dāneshname-ye ʿAlāʾī, Ṭabīʿāt*, ed. Moḥammad Meshkāt. Tehran: Enteshārāt-e anjoma-e āthār-e millī.

Ibn Sīnā. 1970. *Avicenna's De anima (Arabic text): Being the Psychological Part of Kitāb al-Shifāʾ*, ed. Fazlur Rahman. London: Oxford University Press.

Ibn Sīnā. 1970. *Al-Shifāʾ, al-Manṭiq, al-ʿIbāra*, ed. Maḥmūd al-Khuḍayrī. Cairo: Dār al-Kitāb al-sharq.

[59] He therefore strongly rejects the idea that astrology could provide this kind of knowledge. See the final part of Chap. 1, Book 10, of the *Ilāhiyyāt* of the *Shifāʾ*.

Ibn Sīnā. 1974. *Kitāb al-Hidāya li-Ibn Sīnā*, ed. Muḥammad ʿAbduh. Cairo: Maktabat al-Qāhira al-Ḥāditha.
Ibn Sīnā. 1984. *Al-Mabdaʾ wa-l-maʿād*, ed. ʿAbdallāh Nūrānī. Tehran: McGill University in Collaboration with Tehran University.
Ibn Sīnā. 1384/1985. *Al-Najāt min al-gharq fī baḥr al-ḍalālāt*, ed. Moḥammad Taqī Dānishpazhūh. Tehran: Dāneshgāh-e Tehran.
Ibn Sīnā. 1988. Fī Aqsām al-ʿulūm al-ʿaqliyya. In *Al-Madhhab al-tarbawī ʿinda Ibn Sīnā*, ed. ʿAbdalamīr Shamsaddīn, 261–272. Beirut: al-Shirka al-ʿālamiyya li-l-kitāb.
Ibn Sīnā. 2018. *Al-Qānūn fī l-ṭibb*, vol. 1: *Fī al-ʿumūr al-kulliyya min ʿilm al-ṭibb*, ed. Najafgholi Habibi. Hamadan, Iran: International Avicenna Scientific and Cultural Foundation.
Landauer, Samuel. 1875. Die Psychologie des Ibn Sînâ. *Zeitschrift der Deutschen Morgenländischen Gesellschaft* 29: 335–418.
Lizzini, Olga. 1995. La Metafisica del *Libro della guida*: Presentazione e traduzione della terza parte (*bâb*) del *Kitâb al-Hidâya* di Avicenna. *Le Muséon* 108: 367–424.
Özcan, Ahmet. 1993. İbn Sînâ'nın el-Hikmetu'l-Meşrikiyye adlı eseri ve tabiat felsefesi. Ph.D. diss., Marmara Üniversitesi (İstanbul) [Edition of the Natural parts of the *Mashriqiyyūn* (*Easterners*)].
Shah, Mazhar H. 1966. *The General Principles of Avicenna's* Canon of Medicine. Karachi: Naveed Clinic [English translation of Book 1].
al-Ṭūsī, Nasīr al-Dīn. 2007. *Sharḥ al-Ishārāt wa-l-tanbīhāt, al-juz' al-thānī min al-ḥikma*, ed. Ḥasanzādeh al-Āmolī. Qom: Būstān-e Kitāb.
Yusuf Ali, A. 1968. *The Holy Quran: Text, Translation and Commentary*. Beirut: Dar al-Arabia, 1968.

Secondary Literature

Albertini, Tamara. 2010. Dreams, Visions, and Nightmares in Islam: From the Prophet Muhammad to the Fundamentalist Mindset. In *Presenting the Past: Central Issues in Medieval and Early Modern Studies across the Disciplines*, ed. Nancy van Deusen, vol. 2, 167–182. Leiden: Brill.
Bennet, David. 2022. Avicenna's Dreaming in Context. In *Forms of Representation in the Aristotelian Tradition*, vol. 2: *Dreaming*, eds. Christina Thomsen Thörnqvist and Juhana Toivanen, 88–109. Leiden: Brill.
Fahd, Toufic. 1996. *La divination arabe: Études religieuses, sociologiques et folkloriques sur le milieu natif de l'Islam*. Leiden: Brill.
Gätje, Helmut. 1971. *Studien zur Überlieferung des aristotelischen Psychologie im Islam*. Heidelberg: Carl Winter.
Gutas, Dimitri. 2006. Imagination and Transcendental Knowledge in Avicenna. In *Arabic Theology, Arabic Philosophy: From the Many to the One*, ed. James E. Montgomery, 337–354. Leuven: Peeters en Departement Oosterse Studies.
Gutas, Dimitri. 2014. *Avicenna and the Aristotelian Tradition: Introduction to Reading Avicenna's Philosophical Works*. 2nd, revised edition. Leiden: Brill.
Gutas, Dimitri. 2014. The Empiricism of Avicenna. In Dimitri Gutas, *Orientations of Avicenna's Philosophy: Essays on his Life, Method, Heritage*, Chap. 7. Farnham, UK: Ashgate.
Gutas, Dimitri. 2014. Intellect Without Limits: The Absence of Mysticism in Avicenna. In Dimitri Gutas, *Orientations of Avicenna's Philosophy: Essays on His Life, Method, Heritage*, Chap. 12. Farnham, UK: Ashgate.
Hansberger, Rotraud. 2008. How Aristotle Came to Believe in God-Given Dreams. In *Dreaming across Boundaries: The Interpretation of Dreams in Islamic Lands*, ed. Louise Marlow, 50–77. Cambridge, MA: Harvard University Press.
Hansberger, Rotraud. 2018. Representation of Which Reality? 'Spiritual Forms' and 'maʿānī' in the Arabic Adaptation of Aristotle's Parva Naturalia. In *The Parva naturalia in Greek, Arabic*

and Latin Aristotelianism: Supplementing the Science of the Soul, ed. Börje Bydén and Filip Radovic, 99–121. Cham: Springer.

Hasse, Dag Nikolaus. 2000. *Avicenna's De anima in the Latin West: The Formation of a Peripatetic Philosophy of the Soul, 1160–1300*. London: The Warburg Institute.

Janssens, Jules. 2004. 'Experience' *(tajriba)* in Classical Arabic Philosophy (al-Fārābī–Ibn Sīnā). *Quaestio* 4: 45–62.

Janssens, Jules. 2017. Ibn Sīnā on 'Experience' (*tajriba*) and Its Particular Significance for the Evaluation of Simple Drugs. In *Al-Mu'tamar al-ʿalamī al-awwal li-tārīkh al-ʿulūm al-taṭbiqiyya wa-l-ṭibbiyya ʿinda l-ʿArab wa-l-muslimīn*, vol. 5, 299–309. Riadh: Jāmiʿat al-Imām Muḥammad ibn Saʿūd al-Islāmiyya.

Lane, Edward William. 1874. *An Arabic-English Lexicon*. London: Williams and Norgate.

Van Nuffelen, Peter. 2014. Galen, Divination and the Status of Medicine. *Classical Quarterly* 64 (1): 337–352.

Jules Janssens Collaborator at the De Wulf-Mansion Centre, KU Leuven, associated researcher at CNRS, UMR 8230, and director of the "Avicenna latinus" project (UAI) since 2022, Jules Janssens is the author of many books and articles, especially on Ibn Sīnā.

Chapter 4
Translating Epistemic Norms into Social Hierarchy: The Social and Emotional Repercussions of a Theological Controversy

Nimrod Hurvitz

Abstract This study examines the contested underlying assumptions of premodern Islamic theology and their social and emotional repercussions. On one side of this controversy were the speculative theologians or *mutakallimūn*, who asserted that knowledge about the world that is attained through the senses should be part of doing the science of theology because it reveals aspects of God. They also contended that it is possible to attain insights regarding God through speculation based on human rational faculties. Their ideological opponents were the Traditionalists or *muḥaddithūn*, who insisted on a separation between natural scientific knowledge and theology. The Traditionalists argued that human rational faculties, being imperfect, are inadequate for theological investigation; Muslim theological doctrine must derive solely from clear statements that appear in the holy texts. Drawing upon their contrasting views regarding theology, the two currents envisioned very different socio-religious hierarchies. Furthermore, they developed a strong enmity towards each other. When these three factors—contrasting perceptions of theology, conflicting socio-religious hierarchies, and mutual contempt—are taken into consideration, it is possible to better understand the tension and violence that erupted between them throughout the premodern period.

Keywords *Mutakallimūn* · *Kalām* · Traditionalists · Al-Jāḥiẓ · Epistemic norms · Emotions · Social hierarchy

---

N. Hurvitz (✉)
Department of Middle East Studies, Ben Gurion University, Beersheba, Israel
e-mail: nhurvitz@bgu.ac.il

H. C. Erlwein and K. Krause, *Revisiting Premodern Islamic Science and Experience*, SpringerBriefs in History of Science and Technology,
https://doi.org/10.1007/978-3-031-76085-3_4

4.1 Introduction

The Inquisition (*miḥna*) was a pivotal moment in the religious and political history of Islam. It began in AH 218/833 CE when the caliph al-Maʾmūn (r. 198/813–218/833) sent five letters to his governor in Baghdad, ordering him to impose upon his subjects a theological doctrine, and it ended in 233/848 on the orders of the caliph al-Mutawakkil (r. 233/847–247/861). At its outset, the Inquisition seemed likely to succeed. In the first letter, al-Maʾmūn instructed the governor to interrogate "the judges and traditionists" (*al-quḍāt wa'l-muḥaddithīn*) about their position regarding the createdness of the Qur'an by God.[1] Most of them collaborated with the ruler and answered that it was created, though several refused and were imprisoned, and a handful were tortured or died in jail.

Not least because of the forceful intervention of three consecutive caliphs in the articulation of theological doctrine, this moment in Islamic history has been studied in some detail. Much attention has been paid to al-Maʾmūn's motivation, asking whether he was supporting specific theological movements such as the Shiʿis or the Muʿtazilīs or using a theological problem to assert caliphal authority.[2] The literature has also addressed stances on the role of reason in theological inquiries: its endorsement by speculative theologians (*mutakallimūn*) and its rejection by Traditionalists (*muḥaddithūn*).[3]

This chapter investigates the Inquisition from a different angle. It highlights that the hostility and mutual rejection of the *mutakallimūn* and *muḥaddithūn* characteristic of this episode arose from contrasting epistemic norms concerning how the science of theology should be pursued. "Epistemic norms," writes Peter J. Graham, "govern what we ought to say, do, or think from an *epistemic point of view*, from the point of view of promoting true belief and avoiding error." To understand them, Graham continues, one will need to inquire about what is thought to constitute "relevant

[1] Al-Ṭabarī, *Taʾrīkh*, 3:1112; Bosworth (trans.), *The Reunification*, 199. The term *muḥaddithūn* is translated in the rest of this chapter as "Traditionalist"; Bosworth uses the alternative translation, "traditionist," as does Burnett in this volume. Here, I follow Abrahamov's depiction of the *muḥaddithūn* as "those who have regarded religious knowledge as deriving from the Revelation (the Qur'an), the Tradition (the Sunna) and the Consensus (*ijmāʿ*) and preferred these sources to reason in treating religious matters." Abrahamov, *Islamic Theology*, ix. Patricia Crone uses the same term, emphasizing that it is not a good translation but is in wide use. Crone, *God's Rule*, 125–126.

[2] For a historiographical survey addressing these three hypotheses to explain al-Maʾmūn's motives, see Nawas, "Reexamination."

[3] On the cleavage within the Sunni community between theologians and Traditionalists regarding theological speculation, see Schmidtke, "Introduction," 2. For other references in the *Handbook* that address the contrasting views of the *muḥaddithūn* and *mutakallimūn* regarding theology, see Thiele, "Between Cordoba and Nīsābūr," 227–228; Abrahamov, "Scripturalist and Traditionalist," 270–271; El-Rouayheb, "Theology and Logic," 411; Hoover, "Ḥanbali Theology," 625–627. A study that focuses on the contrasting approaches to theology is Abrahamov, *Islamic Theology*. For a discussion of this tension in a philosophical context, see Endress, "The Circle of al-Kindi," esp. 66, and in its political context, Crone, *God's Rule*, 127. On rejection of reason in theological inquiries by Traditionalists, see Abrahamov, "Scripturalist and Traditionalist," 270–272; Hoover, "Ḥanbali Theology," 627–628; Stroumsa, "Early Muslim and Jewish *Kalām*," 216–217.

evidence" and what is regarded as the correct way "to examine and reflect on" such evidence.[4] Applying these observations to the historical context of the Inquisition, this chapter investigates what the theologians and the Traditionalists considered to be relevant and legitimate sources of knowledge for the science of theology and what they considered the right kind of cognitive practices to utilize in articulating the tenets of faith.

Al-Maʾmūn's letter, which endorses the theologians and denigrates the Traditionalists, can be read as indicative of contrasting epistemic norms. Throughout the letter it is evident how these two dominant currents of thought in third/ninth-century Muslim societies viewed the science of theology. For instance, the qualities that the *mutakallimūn* required of participants in theological discourse can be inferred from al-Maʾmūn's criticism of his lay subjects: the "base elements" of the population, he writes, "have no farsightedness, or vision, or faculty of reasoning […] or faculty of seeking illumination by means of the light of knowledge and God's decisive proofs."[5] This "is because of the feebleness of their judgement, the deficiency of their intellects, and the lack of facility in reflecting upon things and calling them to mind."[6] By listing what the "base elements" lack, al-Maʾmūn indicates the qualifications that are needed to engage in the science of theology and that, according to him, only the theologians possess: "faculty of reasoning" (*istidlāl*), "intellect" (*ʿuqūl*), and "knowledge" (*ʿilm*).

The Traditionalists' outlook, too, can be derived from al-Maʾmūn's letter. For example, he mockingly claims that they believe they can take part in theological discourse simply because they are "adherents of the Sunnah [i.e., sayings and doings of the Prophet]" and lead "an ascetic life."[7] His sardonic remarks reveal, by way of criticism, the qualifications that the Traditionalists thought were the prerequisite for engaging in theology: mastery of sacred sources and devoutness.

Al-Maʾmūn's letter is not the only source that addresses the contrasting epistemic norms of theologians and Traditionalists. His accusation that the Traditionalists dissociate human reasoning from theology is echoed in the writings of al-Jāḥiẓ (d. 254/868), a well-known *mutakallim* and belletrist during al-Maʾmūn's reign:[8]

> And the Traditionalists and the masses [*al-ʿawāmm*] are those who adopt [ideas] without questioning [*yuqallidūn*] and do not infer [*lā yuḥaṣṣilūn*] and they do not choose, [yet] unquestioning adoption [*al-taqlīd*] is objectionable when dealing with rational reasoning, and forbidden by the Qur'an [*manhī ʿanhu fī al-Qurʾān*].[9]

In this juxtaposition of the methods of the Traditionalists and the *mutakallimūn*, al-Jāḥiẓ specifies exactly how they differ: whereas the *mutakallimūn* rely on inference (*yuḥaṣṣilūn*) and acts of choosing, the *muḥaddithūn* rely on unquestioning adoption

---

4 Graham, "Epistemic Normativity," 247.

5 Al-Ṭabarī, *Taʾrīkh*, 3:1113; Bosworth (trans.), *The Reunification*, 200.

6 Al-Ṭabarī, *Taʾrīkh*, 3:1113; Bosworth (trans.), *The Reunification*, 200–201.

7 Al-Ṭabarī, *Taʾrīkh*, 3:1114; Bosworth (trans.), *The Reunification*, 201–202.

8 On al-Jāḥiẓ and his thought, see Montgomery, *Al-Jāḥiẓ*, 4–5.

9 Al-Jāḥiẓ, *Rasāʾil*, 3:298.

(*yuqallidūn*). Inference and choice require expertise and the ability to extrapolate insights from texts, information contained in them, and nature. By contrast, *taqlīd* (unquestioning adoption) relies solely on the literal meaning of sacred texts and, by implication, a rejection of human reasoning when inquiring about the divine. Al-Jāḥiẓ ends his criticism of the Traditionalists by pointing out that the Qur'an—the ultimate religious authority in the eyes of the Traditionalists—prohibits unquestioning adoption in this context.

Al-Maʾmūn's letter and al-Jāḥiẓ's account clearly lean towards the theologians when discussing ways of doing theology, but Traditionalist sources, too, have much to tell us about the contrasting epistemic norms of the two groups. Interestingly, some of the Traditionalists agree with the description given by al-Jāḥiẓ. In an account of the *miḥna* interrogation, Ḥanbal b. Isḥāq (d. 273/886) cites one of the leading figures in the *muḥaddithūn* camp, Aḥmad b. Ḥanbal (d. 241/855), who argues that his fellow Traditionalists rightly refuse to apply rational faculties to theological inquiry:

> O Commander of the Faithful, they [the *mutakallimūn*] lack means of discrimination or elucidation on this matter, and the knowledge to which they [the *mutakallimūn* interrogators] summon me is not of the Book of God nor the *sunna* of his Prophet, [but] interpretation which they interpreted and [personal] view which they held, and the Prophet prohibited disputation regarding the Qur'an, and said: "Dispute about the Qur'an is unbelief [*kufr*]," and I am not an adherent of dispute nor of *kalām*, I am solely an adherent of traditions [*āthār wa-akhbār*].[10]

Both sides, it can be seen, agreed that Ibn Ḥanbal and many of the Traditionalists were not conversant with theology (*kalām*)—but they disagreed about the implications of that fact. From Ibn Ḥanbal's point of view, stating "I am not an adherent of dispute nor of *kalām*" means that *kalām* is not relevant to the articulation of Muslim dogma, which is why he ignores this intellectual approach and body of knowledge. It is an act of rejection, not an admission of intellectual limitations. From the *mutakallimūn*'s point of view, in contrast, it is evidence that the *muḥaddithūn* do not have the intellectual abilities to understand theology, and therefore they should not partake in theological discourse.

In this chapter, I first present in more detail the opposing epistemic norms held by theologians and Traditionalists, and highlight three cognitive practices that evolved out of these norms. I then show that these conflicting outlooks had far-reaching social and emotional repercussions. Finally, I contend that conflicting views pertaining to epistemic norms were translated into hierarchies of knowledge and religious authority and, coupled with mutual animosities, generated centuries of intellectual controversies and outright violence.

---

[10] Ḥanbal, *Dhikr*, 60.

4.2 Constructing Epistemic Norms

In the Islamic context, the two questions of epistemic normativity highlighted by Graham—the nature of evidence and the mode of reflection upon it—can be articulated in the following manner:

1. Nature of Evidence: Should epistemic frameworks, which are derived in great part from texts that were written before the rise of Islam in foreign cultures (*ʿulūm al-awāʾil*) and translated into Arabic, be relied upon when discussing theological issues? And should empirical evidence, as suggested in the same sources, be relied upon to solve theological questions?
2. Mode of Reflection: Should modes of rational reasoning be relied upon when seeking definitive answers to theological questions, or are other modes, such as indiscriminating acceptance of sacred texts, preferable?

These two areas of epistemic normativity and the questions that revolved around them elicited a variety of answers. They gave rise to two major theological trends, the Traditionalists and *mutakallimūn*, who contradicted one another.

Let us begin with the first aspect, what counts as evidence. From the *mutakallimūn*'s point of view, large parts of the "relevant evidence" were translations of scientific writings that had been composed in non-Islamic societies. This translation project probably began as early as the last decades of the Umayyad rule (early second/eighth century) and intensified during the early Abbasid era (mid-third/ninth century).[11] The texts that were translated covered a wide variety of disciplines. Some, such as botany, zoology, and medicine, leaned heavily on sense perception and experience; others, such as arithmetic, logic, and metaphysics, were of a more speculative nature.[12] Historians studying the translation movement, such as Dimitri Gutas, have emphasized that the people who took an interest in these texts were not merely a thin layer of detached scientists but rather "the entire elite of ʿAbbasid society: caliphs and princes, civil servants and military leaders, merchants and bankers, scholars and scientists."[13] Furthermore, Gutas notes that the translation movement "percolated downwards to the literate but not so affluent strata of the population." This observation accords with the account given by the Abbasid historian al-Masʿūdī (d. 345/956), who wrote that after the works were translated during al-Manṣūr's reign (d. 158/775), "they were made public, [and] the people read and studied them avidly."[14] The knowledge found in these scientific texts became part and parcel of Abbasid society, stimulated further research that widened the scope of Muslim scholars' knowledge,

[11] On this chronology and the project's motivation, see Saliba, *Islamic Science*, 16. For a somewhat different chronology, emphasizing the early Abbasid era, see Gutas, *Greek Thought*, 2.

[12] Gutas, *Greek Thought*, 1.

[13] Gutas, *Greek Thought*, 2, and 107–150 for a detailed discussion of the social strata that encouraged and used this knowledge. On its impact on society, see also Saliba, *Islamic Science*, 76–77.

[14] Al-Masʿūdī, *Murūj al-dhahab*, 4:388; see Gutas, *Greek Thought*, 135.

served officials in various capacities, and offered scientific knowledge to physicians, veterinarians, and other professionals.[15]

The new empirical knowledge about nature and medicine also shifted the focus of the theological discourse to a new set of questions. As Gutas observes, whereas during the first century of Islam, theological debates "were the result of political and social developments," from the end of Islam's second century the spotlight of theological discourse moved "to subjects which forced the opposing debating groups to have recourse to translated sources."[16] From the early Abbasid era, the theological discourse of the *mutakallimūn* was permeated with conceptualizations and vocabulary based upon their understanding of nature. This point is made succinctly by Gutas: the "abstract theological discussions we hear about have to do with questions of physical theory—atoms, space, and the void."[17] The interplay between natural science and the science of theology has been noted by several scholars in these fields, such as Josef van Ess, who commented that atomism, for example, "made it clear how divine will worked."[18]

The interconnectedness of "foreign sciences" (*ʿulūm al-awāʾil*) translated into Arabic and the *mutakallimūn*'s theology runs across several topics.[19] As well as playing an important role in discussions of God's attributes, it also surfaces in discussions of occasionalism, which asks about "God's absolute power" and articulates theories about forms of causality in nature.[20] The mainstream Sunni *mutakallimūn* argued that God is the sole causal power in the world, hence "negating any kind of natural causality and attributing every causal effect in the world immediately to Him."[21] The effort to comprehend causality in the world compels the believer to bring together God and nature, and in so doing to bring the knowledge attained through the foreign sciences to bear on theological questions such as divine omnipotence. As a result of the need to know about nature when addressing questions about God, "the physical structure of this world became one of the central topics in third/ninth-century Islamic theology."[22] The intertwined nature of physics and God, illustrated by the discussion of occasionalism, highlights one of the epistemic norms espoused by the *mutakallimūn*—that theological investigations should draw upon natural science.

Yet despite the enthusiastic reception of foreign sciences, and the crucial role they came to play in Muslim societies, they also prompted disagreements and competition. According to George Saliba, the Islamic intellectual and administrative milieu was divided into two main currents. One comprised "those who possessed the information contained in the 'foreign sciences'"; they were "employed at the highest echelons of the government offices." The second group "were those who possessed the mastery

---

[15] On the naturalization of scientific knowledge, see Sabra, "Appropriation."

[16] Gutas, *Greek Thought*, 70.

[17] Gutas, *Greek Thought*, 71. On this point, Gutas draws on Dhanani, *Physical Theory*, 182–187.

[18] Ess, *Theology and Society*, 515.

[19] Crone, "Excursus II," 103.

[20] Rudolph, "Occasionalism," 347. On attributes, see Thiele, "Abū Hāshim al-Jubbāʾī."

[21] Rudolph, "Occasionalism," 347.

[22] Rudolph, "Occasionalism," 349.

of the Arabic language and who worked at the lower echelons of the government at the old *dīwān* jobs but now allied to religious figures."[23] The latter allied themselves "with those who steered away from the proximity of political power, in opposition to those who kept on translating 'foreign' sciences into Arabic."[24] The knowledge that was introduced through translations of foreign sciences, leaving its mark on theological discourse, courtly discourse, and the type of individuals who succeeded in the caliphal court, created two currents among the administrators: one that supported the translation movement and one that opposed it. The main factor that determined their attitude towards it was the extent to which it served or hindered their cultural and political interests.[25]

At the far end of the spectrum of religious figures who kept their distance from foreign sciences were conservatives who considered abuse of scientific knowledge a danger that can mislead believers.[26] It should be emphasized that even these conservative thinkers did not regard scientific knowledge in and of itself as unacceptable or as a body of knowledge that ought to be removed from Muslim societies. They did, however, warn against two moves often associated with individuals who studied the sciences:

1. Inserting knowledge about the world into theological discourse or drawing on worldly epistemic frameworks for a discourse about the supra-natural, that is, God. Put differently: applying natural scientific knowledge to answer theological questions.
2. Creating a milieu that belittles evidence appearing in religious texts and places it on a lower rung of epistemic authority than the natural scientific knowledge that they investigate.

Three examples illustrate how conservative Traditionalists depicted the epistemically dangerous marriage between natural scientific and theological scientific knowledge.

The first is a remark made by the historian and jurist al-Subkī (d. 756/1355) about the instigation of the Inquisition. According to al-Subkī, al-Maʾmūn "was among those who took an interest in philosophy and foreign sciences [*ʿulūm al-awāʾil*], excelled in these fields, and conferred with groups of their scholars and this moved him to the doctrine of the created Qur'an."[27] Here, the link between ancient sciences and theology is made quite explicitly. Furthermore, it was not merely an intellectual and theoretical connection. Rather, it was embedded in a social and political context, in which there were two social currents, each of which distrusted and disrespected the other. Finally, as al-Subkī points out, in the specific case of the disagreement over the createdness of the Qur'an, the intellectual controversy led to political persecution.

---

[23] Saliba, *Islamic Science*, 76.

[24] Saliba, *Islamic Science*, 77.

[25] Saliba, *Islamic Science*, 76–77.

[26] This group was studied by Goldziher in a classic but controversial article, Goldziher, "The Attitude." For criticism of some of Goldziher's assumptions, see Gutas, *Greek Thought*, 166–175.

[27] Al-Subkī, *Ṭabaqāt al-Shāfiʿiyya*, 2:56.

The second example of a criticism of conflation between the natural sciences and theology appears in a treatise composed by al-Suyūṭī (d. 911/1505), in which he quotes Abū al-ʿAbbās b. Surayj (d. 306/918). Ibn Surayj, an important Shāfiʿī jurist, commented as follows about the affirmation of God's unity (*tawḥīd*): "*Tawḥīd* of the Muslim people of knowledge and unity is to testify that there is no god but God, and to testify that Muḥammad is his messenger." However, "*tawḥīd* for the people of falsehood is entering into questions of accidents and bodies."[28] In this terse remark, Ibn Surayj criticizes the epistemic link made between the theological science, which investigates God's unity, and the natural sciences, which couch all their discussions in an epistemic framework of "questions of accidents and bodies." He also depicts two clashing currents, "people of knowledge" and "people of falsehood." Here again, there is a clear awareness that society is divided into two camps, one that relies on natural scientific knowledge when discussing theology and another that opposes such integration.

My third example speaks to the second point, the kinds of evidence, and refers to a milieu that places the authority of natural scientific knowledge over that of holy texts. It appears in one of the works of the ninth-century scholar Ibn Qutayba (d. 276/889). In his polemic against the *mutakallimūn*, Ibn Qutayba ascribes to the heretics (*zanādiqa*) and certain philosophers, whom he asserts belong to materialist and atheist trends (*dahriyya*), the view that knowledge gained through the senses is more reliable than revelation. This accusation is made when he writes about individuals who dispute the existence of the *jinn* (invisible creatures mentioned in the Qur'an) and *shayṭān* (the devil): one holding that view "does not believe anything except for what his eyes see, and his senses perceive."[29] Although the supremacy that the individuals labeled *dahriyya* ascribed to sensory perceptions was rejected by most Muslims, including the *mutakallimūn*, it can shed light on the sensitivity of extreme conservative Traditionalists, who considered the foreign sciences a threat to Islamic dogma.

The discussion up until this point reveals that during the third/ninth and fourth/tenth centuries, the *mutakallimūn* and Traditionalists disagreed over the basic epistemic norm of what constitutes evidence in theological discourse. Whereas the *mutakallimūn* took sense perception to be a valid source, together with reason, and argued that experience based on direct sense perceptions plays an important role in investigations about God, the conservative Traditionalists rejected this premise and considered knowledge attained through the senses and speculation to be irrelevant or even dangerous.

Having investigated the first epistemic norm, concerning the nature of evidence, let us now turn to the second, concerning the mode of reflection, on which the *mutakallimūn* and conservative Sunni scholars also disagreed. The bone of contention

---

[28] Al-Suyūṭī, *Ṣawn al-manṭiq*, 118; see also 99–100, 105, 114, 120. For a somewhat unlikely-sounding anecdote about the introduction of the ancient sciences in Muslim society, see 44–45, where the *ṣaḥāba* (Companions of the Prophet) are cast in the role of preventing such sciences from entering Islamic culture and al-Maʾmūn as relying on them, thus damaging the faith.

[29] Ibn Qutayba, *Taʾwīl mukhtalif*, 267; see a similar remark at 429–430.

was how to examine or reflect upon questions pertaining to God, or more specifically, on the place that should be accorded to human reason in the Islamic theological enterprise. Whereas the *mutakallimūn* asserted that inference and interpretation was an indispensable means for theological investigation, Traditionalists downplayed its importance, considering it far inferior to revelation or even denying its relevance altogether. An example of such rejection appears in the Ḥanbalī biographical dictionary known as *Ṭabaqāt al-Ḥanābila* (The Generations of the Ḥanbalīs), where Ibn Ḥanbal is quoted as saying: "Analogical reasoning [*qiyās*] in matters of creed [*dīn*] is invalid."[30] In another, more forceful statement about the incongruence between human reasoning and theological investigation, Ibn Qutayba writes: "As for *kalām*, it is not our business. I do not see [anything that causes] more [people] to perish [i.e., apostatize] than this [*kalām*] and what analogical reasoning imposes upon creed."[31] In this rebuke, Ibn Qutayba is probably referring to the numerous individuals who were lured by a misleading self-confidence in human reasoning and consequently deviated from true faith.

One of the most emphatic critiques of *kalām* and the use of human reason when grappling with theological questions was made by the fourth/tenth-century Ḥanbalī leader, al-Barbahārī (d. 329/941): "And a thought [*wa'l-fikra fī*] about Allah is innovation [*bid'a*]. The Prophet said: 'Think about the created and do not think about Allah.' A thought about the Lord inserts doubt in the heart."[32] It is important to note that al-Barbahārī did not rebuke thinking altogether; he considered the act of thinking about the world—"the created"—to be legitimate and sanctioned by the Prophetic *ḥadīth*. However, thinking about God Himself, this being what the *mutakallimūn* were pursuing when they investigated God's attributes, is forbidden. Al-Barbahārī reiterates this elsewhere, arguing that when it comes to creed, instead of thinking, believers ought to practice *taqlīd*, unquestioning adoption: "Know that creed [*dīn*] is solely *taqlīd*. And *taqlīd* of the Companions of the Prophet."[33] Reasoning about the divine generates disagreements and doubt, both of which are unacceptable in the realm of theology. The only way to avoid this is to accept the authority of the sacred texts or the views of the Companions of the Prophet, "without asking how" (*bi-lā kayfa*).[34]

Despite their insistence on *taqlīd* and *bi-lā kayfa*, conservative theologians were continually challenged by thorny theological issues, and sought a way of tackling them in a manner consistent with their values. That involved basing the discussion about God solely on the Qur'an and *ḥadīth*. The details of this approach have been studied by Livnat Holtzman, writing about a sub-genre of Prophetic traditions, called *aḥādīth al-ṣifāt* (reports about divine attributes), that addresses such issues as "God's place in the universe, His bodily organs, the dialogue that He conducts with humans

---

[30] Ibn Abī Ya'lā, *Ṭabaqāt al-Ḥanābila*, 1:31, and a similar point at 1:241.

[31] Ibn Qutayba, *al-Ikhtilāf fī al-lafẓ*, 21–22.

[32] Ibn Abī Ya'lā, *Ṭabaqāt al-Ḥanābila*, 2:23.

[33] Ibn Abī Ya'lā, *Ṭabaqāt al-Ḥanābila*, 2:29–30.

[34] On the concept of *bi-lā kayfa*, see Abrahamov, "The *bi-lā kayfa* Doctrine."

and His actions."[35] The *aḥādīth al-ṣifāt*, Holtzman points out, enabled the conservative Traditionalists to articulate theological positions while bypassing rational discussions and debates pertaining to them. Holtzman vividly describes the intensity of the disagreements regarding the *aḥādīth al-ṣifāt*: "After the *miḥna*, *aḥādīth al-ṣifāt* became one of the most powerful icons of Islamic traditionalism. As such, they were constantly attacked by the rationalists (mostly the Muʿtazilites)."[36] The *aḥādīth al-ṣifāt* enable a discourse about God that circumvents speculative theology, and as a result they were criticized by the *mutakallimūn*. Furthermore, they embody the contradictory approaches to theology in which the *mutakallimūn* took reasoning for granted, whereas the Traditionalists adamantly opposed it, relied only on revealed sources, and advocated *taqlīd*.

4.3 Constructing Cognitive Practices

The *mutakallimūn*'s assertion that humans ought to inquire about God through rational means generated three contested cognitive practices. The first, speculative theology, was discussed above. The second pertained to the interpretation of the Qur'an. The third was the socio-intellectual practice of debate (*jadal*).

The controversy over interpretation stemmed from the disagreement over the application of human rational faculties to issues pertaining to God. However, in this case, the dispute was not about God's attributes but rather about His revelations, and in particular the effort to make sense of ambiguous verses (*mutashābihāt*). Whereas the *mutakallimūn* considered it legitimate to interpret obscure verses, and often did so by treating them as metaphors, the Traditionalists considered such attempts illegitimate and insisted on understanding the verses literally. From the Traditionalists' point of view, the only way to make sense of ambiguous Qur'anic verses was by relying on other revelations—other Qur'anic verses or the *sunna*. This view is ascribed to Ibn Ḥanbal: "We consider the *sunna* to be reports about the Prophet, and the *sunna* interprets the Qur'an, it is the evidence regarding the Qur'an, and there is no analogical reasoning [*qiyās*] with regard to the *sunna*."[37] According to Ibn Ḥanbal and other conservative thinkers, revealed statements and words must not be ascribed any meaning other than what they literally mean. Any form of "logical" manipulation of meaning that relies on human reasoning, such as *qiyās*, is unacceptable when interpreting the Qur'an.

Regarding the third cognitive practice, *jadal*, the *mutakallimūn* and the Traditionalists disagreed about the necessity of theological debate and whether it clarifies or obfuscates theological matters. Drawing upon the existing polemical tradition in

---

[35] Holtzman, *Anthropomorphism in Islam*, 15. See also Abrahamov, "Scripturalist and Traditionalist," 270–272; Hoover, "Ḥanbali Theology," 626.

[36] Holtzman, *Anthropomorphism in Islam*, 186–187.

[37] Ibn Abī Yaʿlā, *Ṭabaqāt al-Ḥanābila*, 1:241. For a similar assessment of Ibn Ḥanbal's position regarding *tafsīr* or Qur'anic interpretation, see Birkeland, *Old Muslim Opposition*, 12, 19.

the Middle East, the *mutakallimūn* adopted the practice of public debate relying on rational faculties and critical thinking. As Patricia Crone notes: "One way in which reason came to sit in judgement over religious claims was by disputations, a competitive sport of enormous popularity on both sides of the Euphrates both before and after the rise of Islam."[38] Whereas the *mutakallimūn* considered this practice a useful means for addressing theological questions, the Traditionalists asserted that disagreements between two believers over a theological issue "produce doubt" (*yaqdaḥ al-shakk*). This implied a lack of certainty, and therefore should be left outside the realm of Muslim dogma.[39] In Ḥanbalī texts, the prohibition against debating theological issues was explicit: "Avoid quarrel, dispute, and arguments pertaining to creed."[40] The Ḥanbalīs held that disputations were merely another way of introducing the much-maligned method of reason into theology.[41]

During the third/ninth century, the chasm between the epistemic norms and cognitive practices of the *mutakallimūn* and Traditionalists was so wide that in many cases each side was unable to engage the other in honest and open discussion of theological questions. Their relations deteriorated into conflict, heated emotions, and sometimes political violence.

4.4 Translating Epistemic Norms and Cognitive Practices into Social Hierarchy

Epistemic norms and cognitive practices were part of a wider worldview that comprised a variety of ideals and values, including social attitudes and emotional inclinations. The assertion that Islamic doctrine was based on bodies of knowledge and epistemic norms attained through particular cognitive practices drove scholars who had mastered this knowledge to assert their authority and articulate Islamic dogma. The two main currents, the *mutakallimūn* and the Traditionalists, each confident that it was their own duty to formulate Islamic doctrine, imagined society as being organized in a hierarchy with rungs determined by the kind of knowledge possessed. Both placed themselves at the top of that hierarchy.[42]

Ideas about the hierarchical structure of society, coupled with attitudes of self-esteem and disrespect towards others, generated powerful emotions. Condescension and contempt towards other currents of thought, the fear that they would distort

[38] Crone, "Excursus II," 106. See also Treiger, "Origins of *Kalām*," 29; Stroumsa, "Early Muslim and Jewish *Kalām*," 217.

[39] Ibn Abī Yaʿlā, *Ṭabaqāt al-Ḥanābila*, 2:19, 23. On "uncertainty" in theological discussions, see 2:27, where the term *ḥayra* (uncertainty) is used. For a comprehensive collection of criticism regarding debates, see al-Suyūṭī, *Ṣawn al-manṭiq*.

[40] Ibn Abī Yaʿlā, *Ṭabaqāt al-Ḥanābila*, 1:241.

[41] As stated in the opening section of this chapter, the two are mentioned together in Ibn Ḥanbal's interrogation, Ḥanbal, *Dhikr*, 60, when he remarks: "I am not an adherent of dispute nor of *kalām*."

[42] The vision of societies that are stratified according to knowledge is ancient and goes back to Greek and Sassanian thought. See Marlow, *Hierarchy and Egalitarianism*, 42–90.

Islamic dogma, and anxiety about political oppression and street violence raised the stakes, transforming a sophisticated intellectual disagreement among a handful of scholars into a widespread controversy that erupted time and again throughout the Abbasid era and beyond. Perhaps the best-known such eruption was the Inquisition mentioned at the beginning of this chapter.

In this section, I focus on the vision and emotions of the *mutakallimūn*, best captured in the writings of the well-known and prolific *mutakallim* al-Jāḥiẓ. According to al-Jāḥiẓ, the *mutakallimūn* stand at the head of the social hierarchy because they have mastered skills and cognitive practices, based on rational faculties, that enable them to analyze and distinguish truth from falsehood. At the bottom of the hierarchy are the masses (*ʿawāmm*), led by the Traditionalists, who utilize their rational faculties only to a limited extent, mainly in the context of legal thinking, and discard them completely in theological discussions, since they prefer unquestioning adoption in theological matters.

In the opening pages of *Sināʿat al-kalām* (The Art of Kalām), al-Jāḥiẓ discusses the tension between the masses and the *mutakallimūn*, stressing the latters' "aversion to the blind imitation of the gullible and the rabble."[43] Just as in his reference to the masses' uncritical acceptance of theological truisms in the treatise *Khalq al-Qurʾān* cited above, he here presents a binary division of society, in which the *mutakallimūn* develop critical thinking while the masses follow Traditionalist scholars blindly.[44]

Al-Jāḥiẓ emphasizes the abilities and leadership role of the *mutakallimūn* as decisive for their well-deserved authority in society. He argues that without the theologians' analytical and interpretive expertise, lay believers would be in grave danger. It is only due to the intellectual expertise of the *mutakallimūn* that believers can "distinguish between proof and doubt, evidence and what is imagined to be evidence. Through it [*kalām*] it is possible to distinguish between community and faction, *sunna* and innovation."[45] This expertise is essential for guarding Islam, the integrity of which is threatened by all sorts of forces—whether charlatans, false prophets, members of other religions, or misguided believers—who need to be resisted and contested. In the essay *Tashbīh* (Likening God to Creation), al-Jāḥiẓ makes this explicit: "[I]f it were not for *kalām*, true faith would not exist. We would not stand apart from the heretics, and there would not be a distinction between the valid and invalid, nor between the Prophet and the impostor, and proof and subterfuge would not be distinguished, nor evidence and doubt."[46] It is the *mutakallimūn*, according to al-Jāḥiẓ, who have the knowledge and proficiency to guide straying believers and confront the treacherous enemies of Islam.

As well as the social hierarchy, al-Jāḥiẓ also discusses the intellectual hierarchy in which the *mutakallimūn* stand above the rest of society because of their theological expertise. However, he does not dismiss the masses completely. For example, in one of his references to the socio-religious hierarchy of Islam, the masses are allocated

[43] Al-Jāḥiẓ, *Rasāʾil*, 4:243.

[44] Al-Jāḥiẓ, *Rasāʾil*, 3:298.

[45] Al-Jāḥiẓ, *Rasāʾil*, 4:245.

[46] Al-Jāḥiẓ, *Rasāʾil*, 1:285.

a middle position, which assumes that they possess some kind of relevant religious expertise:

> As for the masses [*ʿāmma*], what they comprehend of faith hinges upon the extent of their rational faculties. These abilities do not reach the intellectual capacities nor the abundance of ideas that are attained by scholars [*ʿulamāʾ*]; nor do they descend to the intellectual weakness of the insane and children.[47]

Here, as in many other comments that al-Jāḥiẓ makes about socio-religious stratification, the place of a group or individual in the social ladder is derived from religious expertise and rational faculties. In this case, al-Jāḥiẓ sketches a ladder with three rungs, placing the masses in the middle, above "children and the insane" and below the "scholars." By placing them above children and the insane, al-Jāḥiẓ is certainly not suggesting that they are true experts, but the remark does indicate that they have some sort of expertise. In a forthright comment about the masses and their spheres of knowledge and ignorance, he contends:

> Say to them [those who ask]: As for what they [i.e., the masses] know, it is the revelation itself, without its interpretation [*taʾwīlihi*] and the whole of the *sharīʿa* [Islamic law] itself, and most of the Prophetic traditions [...]. As for what they are ignorant of, it is the interpretation of the revelation, and explanation of the confused [*al-mujmal*] [verses], and the obscure traditions that members of the elite [*al-khawāṣṣ*] transmit to each other.[48]

The masses, writes al-Jāḥiẓ, know the rudiments of the faith. They are familiar with the revealed texts, Qur'an and Prophetic traditions, but their expertise in the sacred texts is limited and superficial, and they are incapable of making sense of obscure Qur'anic verses or Prophetic traditions. The ability to elucidate opaque verses (*mutashābihāt*) belongs to the intellectual elite, who pass this true expertise from generation to generation.

Al-Jāḥiẓ continues this line of discussion, putting forth a socially oriented division of knowledge into two spheres: knowledge possessed exclusively by the elite, and knowledge shared by both the elite (*khāṣṣa*) and the masses (*ʿawāmm*):

> Information [*khabar*] is divided into two: [One is spheres] of information in which the elite do not have an advantage over the commonality, such as the Prophetic traditions regarding the permitted and the forbidden, and chapters dealing with the judiciary, and divorce, and pilgrimage rites, and sales, and beverages, and expiation, and similar matters.[49]

In contrast to the numerous derogatory remarks that al-Jāḥiẓ made about the masses in many of his writings, in these two remarks he acknowledges that in certain spheres of expertise the masses and elite are on a par, mainly with respect to Qu'ran, Prophetic traditions, and the religious law. However, in both cases al-Jāḥiẓ is quick to add that their mastery of revealed knowledge is very basic, since they are incapable of handling the complications that arise in certain parts of the revealed texts and theology.[50] Therefore, although he mentions the masses' expertise, al-Jāḥiẓ does not

---

47 Al-Jāḥiẓ, *Rasāʾil*, 4:43.

48 Al-Jāḥiẓ, *Rasāʾil*, 4:39.

49 Al-Jāḥiẓ, *Rasāʾil*, 4:39.

50 Al-Jāḥiẓ, *Rasāʾil*, 4:39.

alter the hierarchy in which the *mutakallimūn* remain on the top rung as the true epistemic experts.

4.5 Knowledge, Hierarchy, and Body Metaphors

As well as noting that the masses possess basic forms of expertise, al-Jāḥiẓ discusses the mutual reliance of the elite and the masses. He does this through a metaphor of the body.[51] The discussion begins by reiterating the hierarchical relations between the two groups:

> The masses bear the same relation to the elite [*al-khāṣṣa*] as a man's organs bear to him [...]. Just as the organs do not know the mind's intentions, do not think, and so have no reasons to disobey orders, the masses do not know the leadership's intentions, nor the elite's plans, and do not stray from their orders and decisions.[52]

The assertion that the social elite ought to function like the human mind—that is, perceive reality and give orders to the masses, who are the equivalent of the body's organs receiving orders from the mind—obviously reinforces al-Jāḥiẓ's claim of the elite's superior expertise. His choice of this metaphor of the human body suggests that obedience to the elite is part of the natural scheme of things.

But al-Jāḥiẓ adds another dimension to the mind/organs metaphor by arguing that the relations between the brain/elite and the limbs/masses cannot be reduced to a simple hierarchy: "The elite needs the commonality as the commonality needs the elite; it is the same with the heart and the limbs."[53] The interdependence of the two strata complicates the picture. First, it reiterates that the common people possess some form of expertise; otherwise they would be entirely useless and there would be no need to collaborate with them. Second, the gap of power is not as wide as it may seem—since if neither can exist without the other, it implies that the elite are not omnipotent and depend on the commonality to some extent. Nevertheless, although al-Jāḥiẓ does not consider the commonality to be completely ignorant and concedes that they also have a significant social role, he accuses them of lacking self-reflection and being unaware of the limits of their intellectual expertise. After noting that the commonality is on a par with the elite with regard to legal expertise, al-Jāḥiẓ shifts his focus to theology, which, he complains, the commonality lacks the expertise to understand:

> Another sphere about which the masses are ignorant, yet the rabble [*ḥashw*] do enter it unaware of their weakness [is theology] [...] and when a matter such as conversing about God, and anthropomorphism, and the promise and the threat comes up, they enter the fray. Due to their [intellectual] weakness regarding legal opinions they would not engage in that topic, and they would not enter spheres they know nothing about. However, they do

[51] On the use of this metaphor by al-Fārābī, see Marlow, *Hierarchy and Egalitarianism*, 53, and on its appearance in other texts, 159.

[52] Al-Jāḥiẓ, *Rasāʾil*, 4:36–37; English in Pellat, *The Life*, 78, with minor emendations.

[53] Al-Jāḥiẓ, *Rasāʾil*, 4:38; English in Pellat, *The Life*, 79.

> not refrain from conversations about rectitude and injustice, and they do not disengage from conversations about free choice and innate nature [...] and other matters pertaining to God.[54]

From al-Jāḥiẓ's point of view, the fact that the masses have little to no expertise in theology is not a problem in itself. If the masses had internalized their intellectual weaknesses, admitted that they do not have expertise in theology, and adopted the tenets of faith that the *mutakallimūn* have articulated, he would have considered them upright believers. Yet as his words make abundantly clear, the masses refuse to place the *mutakallimūn* at the top of the spiritual hierarchy and allow them to determine Muslim dogma. Instead, all they do is argue with the *mutakallimūn* and articulate their own tenets of faith, and in doing so, they turn the socio-religious order upside down.

4.6 Trust Versus Rational Faculties

The masses' refusal to accept the authority of the *mutakallimūn* derives from the fact that they base religious expertise on different epistemic norms than do the *mutakallimūn*. As shown, the masses ascribe religious authority to the epistemic norms of trust, respect for moral qualities, and mastery of the Qur'an and *sunna*; they do not embrace the epistemic norms of sense perception, epistemic frameworks, and analytic reasoning as contained in the foreign sciences. As al-Maʾmūn wrote in the Inquisition letters, and as al-Jāḥiẓ wrote in his *Rasāʾil*, the masses' propensity to treat the Traditionalists as religious authorities and ignore the *mutakallimūn* was an outcome of embracing the wrong epistemic norms.

What al-Maʾmūn and al-Jāḥiẓ criticized most severely was the masses' insistence that their leaders, the Traditionalists, can articulate tenets of faith and deserve to be listened to. Al-Maʾmūn lambasts the masses and their leaders in the most contemptuous language in the Inquisition letter mentioned above, focusing his criticism on their claim that knowledge of the sources and devout behavior are the epistemic norms that qualify them to speak expertly about theological issues. In his view, the Traditionalist scholars are intellectual charlatans who lead the gullible masses astray.

Al-Jāḥiẓ, whose point of view was similar, wrote a more evenhanded rebuttal of the masses' insistence on their religious expertise. In his treatise on the createdness of the Qur'an, he describes and disparages their claim to possess religious expertise and thus to be entitled to participate in theological discourse. "As for their statement: Among us are the ascetics and the devout," he replies: "Among the devout of the *Khawārij* alone there are more devout believers than there are among them [common people], despite the small numbers of the *Khawārij*."[55] Asceticism and devoutness are no validators of theological dogma because there are plenty of ascetics and devout believers among apostate groups such as the *Khawārij*, who emerged during the first civil war in the first/seventh century and are remembered for their violent attacks on

[54] Al-Jāḥiẓ, *Rasāʾil*, 4:39–40.

[55] Al-Jāḥiẓ, *Rasāʾil*, 3:298.

all those who did not follow them. To put it differently, the fact that there are devout believers and ascetics among the masses is no proof that their theological beliefs are valid. At the same time, al-Jāḥiẓ's comment accurately reflects the weight that the followers of Traditionalism ascribed to religious devotion and moral stature.

Al-Jāḥiẓ is explicitly critical of the authority given to the ascetics simply because they adopt pious practices. He mocks the notion that these practices, often disgusting or asocial, enable their practitioners to take part in theological discourse: "If preference, leadership, rank, and nobility were commensurate with coarse skin, slovenly appearance, numerous fasts, and preference for alienation and seclusion, then ʿUthmān b. Maẓʿūn would precede Abū Bakr al-Ṣiddīq, and Bilāl b. Rabbāḥ would surpass ʿUthmān b. ʿAffān."[56] Like al-Maʾmūn, al-Jāḥiẓ is disdainful of the way ascetics build religious prestige and use it to reach positions of leadership they do not deserve.

Another argument, similar to the first, that is made by the masses and the Traditionalists is that they have wide support among the masses of the believers, the learned and the devout. Al-Jāḥiẓ replies: "And as for their claim: We have the support of the masses, and the devout, and the jurists, and the Traditionalists. The fact of the matter is that only the sectarians support them."[57] From his perspective, the masses' claims to authority are factually wrong and distorted.

The *mutakallimūn*'s criticism of the masses' leadership, and their insistence that those leaders lack the expertise to partake in theological discourse, illustrates that at this time, theological discourse focused not on concrete theological arguments but on mutual *ad hominem* accusations, challenging expertise and authority.

4.7 Arrogance, Fear, and Dialogue of the Deaf

The *mutakallimūn*'s rejection of the Traditionalists' theological expertise both aroused and reflected strong emotions in the two camps. Some of the harshest outbursts are found, as we have seen, in al-Maʾmūn's *miḥna* letters, in which he expressed his profound disrespect and contempt towards the weak intellect of the Traditionalists.[58] He is also outraged by their success—their ability to attain positions of leadership due to the fact that they "deliberately lead astray the ignorant."[59] Perhaps the most telling indication of his emotional state are the numerous instances of verbal abuse and curses, such as calling the Traditionalists liars.[60] In a subsequent letter, al-Maʾmūn's abuse becomes personal: he accuses one scholar of being an

[56] Al-Jāḥiẓ, *Rasāʾil*, 1:301. ʿUthmān b. Maẓʿūn and Bilāl b. Rabbāḥ were two highly respected Companions of the Prophet who were known for their piety and devotion. Abū Bakr al-Ṣiddīq and ʿUthmān b. ʿAffān were the first and third caliphs respectively.

[57] Al-Jāḥiẓ, *Rasāʾil*, 3:297.

[58] Al-Ṭabarī, *Taʾrīkh*, 3:1113; Bosworth (trans.), *The Reunification*, 200.

[59] Al-Ṭabarī, *Taʾrīkh*, 3:1114; Bosworth (trans.), *The Reunification*, 202.

[60] Al-Ṭabarī, *Taʾrīkh*, 3:1115; Bosworth (trans.), *The Reunification*, 203.

unbeliever who utters falsehoods;[61] another he labels an ignoramus and "a child in intellect,"[62] and a third is mocked for having "defective intelligence."[63] Al-Maʾmūn's caustic language illustrates once again the emotional barriers that arose between the two currents of society.

Al-Jāḥiẓ's essays are also laden with emotion. However, as was discussed above, his attitude to the masses and their Traditionalist leaders is more complex. He mentions that the masses know and understand Islamic law, and that they are part of a system in which the elite and the masses depend on each other. Overall, in his essays the reader comes across a more nuanced depiction of the masses, yet al-Jāḥiẓ's ascription of a few merits to the Traditionalists should not mislead us. Al-Jāḥiẓ was extremely proud of the *mutakallimūn* and contemptuous of anyone who did not belong to their ranks. He writes about *kalām* in the following words:

> [Know that] the practice of *kalām* is a costly jewel, precious gem, and inexhaustible treasure, everlasting, and untiring and unchanging friend; that it is the touchstone of all action, the bridle of all utterance, the scale that discovers sufficiency or deficiency, the filter that reveals the purity or impurity of all things, and upon which all men of knowledge rely.[64]

Along with his flowery rhetoric, al-Jāḥiẓ asserts that *kalām*'s strength is its ability to discern between "sufficiency or deficiency" and "purity or impurity." It is the belief that the competences of *kalām* are the means with which true faith is protected, and the individuals who possess this expertise are its sentinels, that moves him to write "had it not been for *kalām*, God's faith would not have been established," and to assert that the *mutakallimūn* ought to serve as the leaders of the Muslim community.[65]

The other side of the *mutakallimūn*'s pride was arrogance. Imagining themselves standing on the highest rung of the socio-religious ladder, they looked down with scorn upon anyone who did not possess their skills. And they became even more aggressive when those they imagined to be on lower rungs of the hierarchy had the insolence to reject their leadership.[66]

It is important to note that by the early third/ninth century, both sides were behaving aggressively towards each other. In his descriptions of the events that preceded the *miḥna*, al-Jāḥiẓ accuses the masses of going far beyond a rejection of the *mutakallimūn*'s authority and harboring intense enmity towards them. He describes their actions thus: "after they cursed us, they began to show affection, after boycotting *kalām* they began to sit and talk to us, after turning a deaf ear they began to listen, after showing hostility they changed."[67] In this somewhat overly optimistic assessment of the *miḥna*'s impact, he describes a grim past.

---

61 Al-Ṭabarī, *Taʾrīkh*, 3:1126; Bosworth (trans.), *The Reunification*, 215.

62 Al-Ṭabarī, *Taʾrīkh*, 3:1127; Bosworth (trans.), *The Reunification*, 216.

63 Al-Ṭabarī, *Taʾrīkh*, 3:1127; Bosworth (trans.), *The Reunification*, 217.

64 Al-Jāḥiẓ, *Rasāʾil*, 4:244.

65 Al-Jāḥiẓ, *Rasāʾil*, 1:285.

66 On the masses' refusal to accept the spiritual authority of the *mutakallimūn*, see al-Jāḥiẓ, *Rasāʾil*, 1: 289, and similar remarks at 285 and 297.

67 Al-Jāḥiẓ, *Rasāʾil*, 1:288.

The *mutakallimūn*'s reaction to the masses' rejection of their authority appears several times in al-Jāḥiẓ's essays. For example, in several pages in al-Jāḥiẓ's *Maqālat al-ʿUthmāniyya*, he writes contemptuously about the masses who have no idea of theology but refuse to admit it.[68] This belittling of the common people goes hand in hand with a disparagement of their intellectual technique, blind imitation: "The people of *ḥadīth* and the masses follow blindly [*yuqallidūn*] and they do not infer."[69] In the relations between the Traditionalists and the *mutakallimūn*, disagreement over tenets of faith and method came to include insults and humiliations.

The offensive language that was used by al-Maʾmūn and al-Jāḥiẓ formed part of the growing tension between the two currents, which gradually turned into a power struggle.[70] This clash came to include various measures by which each side pressured the other, to the extent that both the caliph al-Maʾmūn and al-Jāḥiẓ wrote of fear experienced by the *mutakallimūn*.[71] In the *miḥna* letter where al-Maʾmūn writes about his subjects in obnoxious language, he also accuses them of frightening the *mutakallimūn*: "They are vessels of ignorance, banners [or: milestones, *aʿlām*] of mendaciousness and the tongue of [the devil] Iblīs, who speaks through his companions and strikes terror into the hearts of his adversaries."[72] From al-Maʾmūn's point of view, the companions of the devil are the *muḥaddithūn*, and their adversaries are the *mutakallimūn*. If we take al-Maʾmūn's description at face value, when the Inquisition was instigated, the *muḥaddithūn* were able to terrorize the *mutakallimūn*.

Al-Maʾmūn's terse reference to the fears of the *mutakallimūn* is reiterated by al-Jāḥiẓ, who also writes about the hostility towards the "scholars of the *mutakallimūn*." In al-Jāḥiẓ's account, the Inquisition was intended to remove their fear of the Traditionalists. He writes: "This [the *mutakallimūn*'s fear] continued until God detracted from their [the *muḥaddithūn*'s] might and decreased from their strength [...] and the Inquisition transformed them."[73] In the two treatises, *Khalq al-Qurʾān* and *Tashbīh*, that were written during the Inquisition, al-Jāḥiẓ speaks openly about these tensions, and the expectation that the Inquisition will bring about a shift in the power relations between the currents—making it possible to establish the *mutakallimūn*'s view of knowledge and enable them to implement the proper social hierarchy and dominate society.

The enmity between the two sides led to mutual accusations of apostasy. In *Khalq al-Qurʾān*, al-Jāḥiẓ recounts how each accused the other of hurling the accusation of

---

68 Al-Jāḥiẓ, *Rasāʾil*, 4:36–43.

69 Al-Jāḥiẓ, *Rasāʾil*, 3:298.

70 The background to their fear is a vague yet frightening sense that the masses are dangerous and unruly. After noting that the masses support the anthropomorphists, al-Jāḥiẓ observes that "the elite do not have power over the masses, and the people of distinction do not have power over the base elements," al-Jāḥiẓ, *Rasāʾil*, 1:283. He also writes about the "intimidation of the scholars of the *mutakallimūn*," al-Jāḥiẓ, *Rasāʾil*, 1:285.

71 On the pressures that were applied by the Traditionalists, see Hurvitz, "Miḥna as Self-Defense"; Hurvitz, "al-Maʾmūn." On the decline in numbers of sectarian Traditionalists, apparently due to the pressures placed on them, see Melchert, "Sectaries."

72 Al-Ṭabarī, *Taʾrīkh*, 3:1115; Bosworth (trans.), *The Reunification*, 203.

73 Al-Jāḥiẓ, *Rasāʾil*, 1:285, 287.

unbelief (*kufr*). Taking the side of the *mutakallimūn*, he reproaches the masses: "It is you who hasten the people to accuse us of unbelief."[74] Setting aside the question of who was right, by the first decades of the third/ninth century, mutual demonization precluded intellectual dialogue.

4.8 Concluding Remarks

As this study has demonstrated, the underlying assumptions of Islamic theology—its epistemic norms and cognitive practices—were contested. This controversy led to the formation of several groups of thinkers, the two central ones being the Traditionalists and the *mutakallimūn*. The differences between these currents come into relief in their answers to two methodological questions:

1. Can scientific knowledge, what we learn about nature, the stars, and humans through our senses, be relied upon when contemplating about God?
2. Can humans rely on logical speculations and debates when searching for insights regarding God?

The *mutakallimūn* answered these questions in the affirmative. According to them, knowledge about the world that is attained through the senses should be part of doing the science of theology because it reveals various aspects of God; at the same time, it is possible to attain insights regarding God through speculation based on human rational faculties. By contrast, the Traditionalists' answer was in the negative. According to them, a separation is required between natural scientific knowledge, which draws upon sense perception, and theology; human rational faculties are inadequate for theological investigation because they are imperfect—due to their limitations, humans cannot fathom God or articulate dogma. The Traditionalists therefore rejected all speculation about God and dismissed the claim that human thought can attain a level of certainty capable of transforming speculations into tenets of faith.

The controversy between these two currents of thought evolved around the notion of certainty, in other words, the question of how human beings can be sure that their beliefs about God are absolutely true. Whereas the *mutakallimūn* asserted that human intellect can attain that level of truth, the Traditionalists denied it and argued that the only way an individual can be completely sure about his beliefs pertaining God is through clear statements that appear in holy texts—only, that is, through God's revelation.

These contradictory epistemic norms and cognitive practices also engendered different ways of imagining society. Although they shared the notion that society is structured as a hierarchy, with men of knowledge at its top, they disagreed about who exactly is at the apex—Traditionalist scholars of Islamic religious sciences, or the *mutakallimūn*. Furthermore, each contended that their current of thought protected Islam from the other current, which jeopardized it.

---

[74] Al-Jāḥiẓ, *Rasāʾil*, 3:296.

Concurrently, each group's view of its own superiority generated pride in itself and condescension, fear, and hostility towards their rivals. These attitudes come across clearly in the writing of al-Maʾmūn and al-Jāḥiẓ, in which they use insulting terms to describe the Traditionalists and their followers, praise the *mutakallimūn*'s intellectual abilities, and in the same texts write about their vulnerabilities and fears. Emotions ran high, and the two sides found it increasingly difficult to maintain a substantive theological exchange.

Instead of developing a shared language that would facilitate theological conversation, they accused each other of misunderstanding the theological endeavor and posing a threat to Islam's belief system. Mutual anger and contempt dominated their rhetoric, as they sought to create new alliances and win supporters that would enable them to shape the "true" version of Islam. In the early third/ninth century, as the Traditionalists succeeded in winning the support of large segments of the common people and the *mutakallimūn* secured that of the caliphal court, this theological dispute led to the eruption of the Inquisition. In subsequent centuries, the controversy over epistemic norms and cognitive practices continued to generate intense power struggles that would involve the masses and caliphal court, and lead to boycotts and violence in the streets.

Bibliography

Primary Sources

Ḥanbal b. Isḥāq. 1977. *Dhikr miḥnat al-imām Aḥmad b. Ḥanbal*, ed. Muḥammad Naghash. Cairo: Maṭbaʿat Dār Nashr al-Thaqāfa.
Ibn Abī Yaʿlā. 1951. *Ṭabaqāt al-Ḥanābila*, ed. M. Ḥ. al-Fiqī, 2 vols. Cairo: Maṭbaʿat al-Sunna al-Muḥammadiyya.
Ibn Qutayba. 1991. *Al-Ikhtilāf fī al-lafẓ wa'l-radd ʿalā al-Jahmiyya wa'l-mushabiha*, ed. ʿUmar b. Maḥmūd. Riyadh: Dār al-Rāya.
Ibn Qutayba. 2009. *Taʾwīl mukhtalif al-ḥadīth*, ed. Salīm b. ʿĪd al-Hilālī. Riyadh: Dār Ibn Qayyim.
al-Jāḥiẓ. 1991. *Rasāʾil al-Jāḥiẓ*, ed. ʿAbd al-Sallam Muḥammad Hārūn, 4 vols. Beirut: Dār al-Jīl.
al-Masʿūdī. 1989. *Murūj al-dhahab wa-maʿādan al-jawhar*, ed. Qāsim al-Shimāʾī al-Rifāʿī, 4 vols. Beirut: Dār al-Qilma.
Pellat, Charles. 1969. *The Life and Works of Jāḥiẓ: Translations of Selected Texts*, translated from the French by D.M. Hawke. Berkeley: University of California Press.
al-Subkī. 1964. *Ṭabaqāt al-Shāfiʿiyya al-kubrā*, ed. M. M. al-Ṭanāhī and ʿA. M. al-Ḥilw, 10 vols. Cairo: Maṭbaʿat ʿIsā al-Bābī al-Ḥalbī.
al-Suyūṭī. 1970. *Ṣawn al-manṭiq wa'l-kalām ʿan fann al-manṭiq wa'l-kalām*, ed. ʿAlī Sāmī al-Nashshār and Suʿād ʿAlī ʿAbd al-Rāzzaq. Cairo: Majmaʿ al-Buḥūth al-Islāmiyya.
al-Ṭabarī. 1987. *The Reunification of the ʿAbbāsid Caliphate*, translated by Clifford E. Bosworth, vol. 32 of *The History of al-Ṭabarī*. Albany: SUNY Press.
al-Ṭabarī. 1879–1901. *Taʾrīkh al-rusul wa'l-mulūk*, ed. M. J. de Goeje et al., 15 vols. Leiden: Brill.

Secondary Literature

Abrahamov, Binyamin. 1995. The *bi-lā kayfa* Doctrine and Its Foundations in Islamic Theology. *Arabica* 42: 365–379.
Abrahamov, Binyamin. 1998. *Islamic Theology: Traditionalism and Rationalism*. Edinburgh: Edinburgh University Press.
Abrahamov, Binyamin. 2016. Scripturalist and Traditionalist Theology. In *The Oxford Handbook of Islamic Theology*, ed. Sabine Schmidtke, 263–279. Oxford: Oxford University Press.
Birkeland, Harris. 1956. *Old Muslim Opposition against Interpretation of the Koran*. Uppsala: Almqvist & Wiksell.
Crone, Patricia. 2016. Excursus II: Ungodly Cosmologies. In *The Oxford Handbook of Islamic Theology*, ed. Sabine Schmidtke, 103–129. Oxford: Oxford University Press.
Crone, Patricia. 2004. *God's Rule: Government and Islam*. New York: Columbia University Press.
Dhanani, Alnoor. 1993. *The Physical Theory of Kalām: Atoms, Space, and Void in Basrian Muʿtazilī Cosmology*. Leiden: Brill.
El-Rouayheb, Khaled. 2016. Theology and Logic. In *The Oxford Handbook of Islamic Theology*, ed. Sabine Schmidtke, 408–431. Oxford: Oxford University Press.
Endress, Gerhard. 1997. The Circle of al-Kindi: Early Arabic Translations from the Greek and the Rise of Islamic Philosophy. In *The Ancient Tradition in Christian and Islamic Hellenism*, ed. Gerhard Endress and Remke Kruk, 43–76. Leiden: Research School CNWS.
Goldziher, Ignaz. 1981. The Attitude of Orthodox Islam Towards the 'Ancient Sciences.' In *Studies in Islam*, edited and translated by Merlin L. Swartz, 185–215. Oxford: Oxford University Press.
Graham, Peter J. 2015. Epistemic Normativity and Social Norms. In *Epistemic Evaluation: Purposeful Epistemology*, ed. David K. Henderson and John Greco, 247–73. Oxford: Oxford University Press.
Gutas, Dimitri. 1998. *Greek Thought, Arabic Culture: The Graeco-Arabic Translation Movement in Baghdad and Early ʿAbbasid Society (2nd–4th/8th–10th Centuries)*. London: Routledge.
Holtzman, Livnat. 2018. *Anthropomorphism in Islam: The Challenge of Traditionalism (700–1350)*. Edinburgh: Edinburgh University Press.
Hoover, Jon. 2016. Ḥanbali Theology. In *The Oxford Handbook of Islamic Theology*, ed. Sabine Schmidtke, 625–646. Oxford: Oxford University Press.
Hurvitz, Nimrod. 2016. Al-Maʾmūn (r. 198/813–218/833) and the *Miḥna*. In *The Oxford Handbook of Islamic Theology*, ed. Sabine Schmidtke, 649–659. Oxford: Oxford University Press.
Hurvitz, Nimrod. 2001. *Miḥna* as Self-Defense. *Studia Islamica* 92: 93–111.
Marlow, Louise. 1997. *Hierarchy and Egalitarianism in Islamic Thought*. Cambridge: Cambridge University Press.
Melchert, Christopher. 1992. Sectaries in the Six Books: Evidence for Their Exclusion from the Sunni Community. *The Muslim World* 82: 287–295.
Montgomery, James. 2013. *Al-Jāḥiẓ in Praise of Books*. Edinburgh: Edinburgh University Press.
Nawas, John A. 1994. A Reexamination of Three Current Explanations for al-Maʾmun's Introduction of the *Miḥna*. *International Journal of Middle East Studies* 26: 615–629.
Rudolph, Ulrich. 2016. Occasionalism. In *The Oxford Handbook of Islamic Theology*, ed. Sabine Schmidtke, 347–363. Oxford: Oxford University Press.
Sabra, Abdelhamid I. 1987. The Appropriation and Subsequent Naturalization of Greek Science in Medieval Islam: Preliminary Statement. *History of Science* 25, 223–243.
Saliba, George. 2011. *Islamic Science and the Making of the European Renaissance*. Cambridge, MA: The MIT Press.
Schmidtke, Sabine. 2016. Introduction. In *The Oxford Handbook of Islamic Theology*, ed. Sabine Schmidtke, 1–23. Oxford: Oxford University Press.
Stroumsa, Sarah. 2019. Early Muslim and Jewish *Kalām*: The Enterprise of Reasoned Discourse. In *Rationalization in Religions, Judaism, Christianity and Islam*, ed. Yohanan Friedmann and Christoph Markschies, 202–223. Berlin: De Gruyter.

Thiele, Jan. 2016. Abū Hāshim al-Jubbā'ī's (d. 321/933) Theory of 'States' (*aḥwāl*) and Its Adaption by Ashʿarite Theologians. In *The Oxford Handbook of Islamic Theology*, ed. Sabine Schmidtke, 364–383. Oxford: Oxford University Press.
Thiele, Jan. 2016. Between Cordoba and Nīsābūr: The Emergence and Consolidation of Ashʿarism (Fourth-Fifth/Tenth-Eleventh Century. In *The Oxford Handbook of Islamic Theology*, ed. Sabine Schmidtke, 225–241. Oxford: Oxford University Press.
Treiger, Alexander. 2016. Origins of *Kalām*. In *The Oxford Handbook of Islamic Theology*, ed. Sabine Schmidtke, 27–43. Oxford: Oxford University Press.
van Ess, Josef. 2018. *Theology and Society in the Second and Third Centuries of the Hijra*, vol. 4. Translated by Gwendolin Goldbloom. Leiden: Brill.

Nimrod Hurvitz teaches at the Department of Middle East Studies, Ben Gurion University. He has published on Muslim law, courts of law, the politics of theology, and modern and medieval socio-religious movements.

Chapter 5
Epilogue: Experiencing Experiences—Four Encounters with Experience in the Medieval Islamicate World

Jon McGinnis

Abstract This epilogue explores and sometimes expands on many of the recurrent themes of this volume. It takes seriously the editors' observation about the subject-rootedness of experience as understood in the medieval Islamicate world and about the notion of internalized objectivity as a lens for reassessing our histories of science. It argues that rather than holding to our current sharp, even arbitrary, divide between the objective and subjective, it is more fruitful to approach Islamicate science as one that views the relation between the experience of the individual scientists and the standards of a given scientific community as forming an ecosystem. Using the interpretive frame set out in the volume's introduction, the epilogue supplements and contextualizes the essays to show that both subject-rootedness and internalized objectivity were operative throughout the scientific endeavors of medieval Islamic science. Against this background, a number of disciplines that have been excluded from or marginalized in our histories of science—such as grammar, theology, astrology, and dream interpretation—emerge as areas of scientific and historical interest.

Keywords Experience · Islamic science · Ibn Sīnā/Avicenna · Al-Sīrāfī–Mattā debate · Epistemic norms · Epistemic cultures

We use the English term "experience" in a number of different ways. Consider the following exchanges:

1. "Describe bourbon?"—"One *experiences* an amber liquid that is slightly spicy, frequently with notes of cinnamon, caramel, and even almond" (a description of the perception of an external object);
2. "What do you *experience* when *you* drink bourbon?"—"Aah, the *experience* is one 'to warm the heart, to reduce the anomie of the late twentieth century, to

J. McGinnis (✉)
University of Toronto, Toronto, Canada
e-mail: jon.mcginnis@utoront.ca

H. C. Erlwein and K. Krause, *Revisiting Premodern Islamic Science and Experience*, SpringerBriefs in History of Science and Technology,
https://doi.org/10.1007/978-3-031-76085-3_5

cut the cold phlegm of Wednesday afternoons'" (an evaluation of an internal perception);[1]
3. "Can you recommend a good bourbon?"—"Yes, I have the *experience* to judge a good bourbon from a bad one" (a kind of knowledge, expertise, or skill);
4. "How did you acquire your knowledge of bourbon?"—"I acquired it from my many *experiences* with bourbon" (the means by which knowledge is acquired).

As these examples suggest, we use "experience" to cover a host of senses: (objective) descriptions, (subjective) evaluations, a kind of knowledge, and even the means to knowledge.

While it may be too much to suggest that "experience" is an equivocal term, it certainly has multifaceted dimensions, aspects, or phases, particularly if one considers its role in science. Experience is sometimes thought of as an *input*, for example the observations that inform and constrain a scientific theory. Sometimes it is viewed as an *output*, for example scientific knowledge, expertise, and practices. Sometimes experience is conceived not merely as a relatum but also as in some way making up the very relation between relata, and so as a *function*, for example the scientific method, for it is experience that allows scientists to input observations and then suggest hypotheses, perform tests, and analyze findings that ultimately output scientific knowledge. One further dimension of experience—and again I am sure there are more—is as *self-reflective* or *evaluative*. In this latter case, one's past experiences determine the weight, significance, or appreciation that one places on some phenomenon or data set so as to inform one's present experience. For instance, Norwood Hanson has one imagine the astronomers Tycho Brahe (d. 1601) and Johannes Kepler (d. 1630) watching the sun's just appearing on the horizon in the early morning. Both Kepler and Brahe experience the same phenomenon; they both experience the sun on the horizon. Brahe, however, experiences the sun *as* rising above the horizon, whereas Kepler experiences the horizon *as* sinking below the sun. "Kepler regarded the sun as fixed: it was the earth that moved," writes Hanson, but "Tycho followed Ptolemy and Aristotle in this much at least: the earth was fixed and all other celestial bodies moved around it."[2] The culmination of these two scientists' own unique experiences—their scientific practices, background knowledge, theoretical assumptions and even aesthetic preferences—led to their adoption of their respective theories and shaped their observations such that they evaluated their experiences of the same phenomenon differently.

In the introduction to this volume, the editors point to several of the multifarious dimensions of experience that historians of the Scientific Revolution and its legacy in the West have used to understand and to explain the history of science. These histories appealed initially to the scientific method, understood as the *empirical* method, and later to scientific practices, where practices are thought of in terms of *empirical pursuits*, subsuming the scientific method among other things. Of course,

[1] For the aesthetics of "knocking [a bourbon] back neat," see Walker Percy's delightful, even if by his own assessment "unedifying," essay "Bourbon."

[2] Hanson, *Patterns of Discovery*, 5.

"empirical," derived from the Greek *empeiria*, simply means connected with or characterized by experience. The presumption is that experience is a good-making feature in both scientific method and practice. What is it, however, that makes experience a good-making feature? What are the epistemic norms and values associated with experience? In other words, what experiences are good, bad, or indifferent relative to science? Are observations gathered either directly or indirectly through the external senses good, while introspection and first-person reports are *bad* or at least perhaps dubious? Is quantitative data, data presented in terms of numbers and measurements, *better* scientifically than qualitative data, data presented in terms of the scientist's own affections and perspectives? In general, does a scientist's reliance on "objectivity" produce *good* scientific experience, while relying on "subjectivity" makes for *bad* (or at least not as good) experience in one's overall evaluation of scientists and their research programs?

For the most part, contemporary scientists and even historians of science have tended to say, "Yay objectivity, boo subjectivity." Yet all of the above questions and the answers that contemporary scientists, philosophers, and historians give, argue the editors, are based upon a potentially artificial—but certainly historically contingent—Kantian dichotomy between objectivity and subjectivity.[3] Subjectivity is frequently understood as personal feelings, biases, emotions and is opposed to objectivity, understood as public rules, standards, criteria, even expectations applicable to all, or at least all who belong to a particular epistemic community, a point to which I return below.

The current volume challenges this dichotomy between "good" objective experience and "bad" subjective experience on two fronts. It asks, and then explores, two questions: (1) How does one's evaluation of experience affect what one counts as a legitimate science worthy of pursuit?, and (2) What is the role of the individual's own personal context, background knowledge, assumptions, values, and preferences in determining what counts as the experience relevant to science?

Concerning the first question, historians of science have focused primarily (although admittedly not exclusively) on *quantitative* sciences, such as medicine, astronomy, optics, and in general natural philosophy to the extent that its data and conclusions are given in terms of measurements, mathematical formula, and the like. Possessing such quantifiable aspects, the narrative goes, makes for legitimate sciences because the data are objective and accessible to all to test, to confirm, and to repeat; quantifiable aspects have purported good-making features that first-person reports and qualitative aspects either purportedly lack or are less public. Because of our current preference for "objectivity," a number of sciences, which were included in the medieval Islamicate scientific curriculum, have either been rejected outright as sciences—notably theology, but also grammar and law—or been marginalized and viewed as pseudoscience, such as astrology and dream interpretation. This rejection and marginalization have resulted from the current distinction between the objective

[3] Cf. Daston and Galison, *Objectivity*, Chap. 4.

and the subjective. Yet if we bracket this distinction, then a host of other sciences—subjects deemed worthy of pursuit within their own historical, scientific context—emerge as worthy of the historian's investigation. Such an investigation is precisely what the present volume offers.

Many contemporary historians of science approach the second question, "What counts as the experience relevant to a science?," in the same way: experience is of scientific value only if it is public, quantifiable, objective, whereas an individual's personal, qualitative, subjective experience is seen as having either no or little scientific value. For instance, even when contemporary scientific inquiries allow first-person reporting of one's experience, they frequently do so only to the extent that such self-reporting can also be rendered quantifiable, and so purportedly objective. Take pain, for example. Pain seems to be as subjective an experience as there is. Yet doctors and nurses no longer ask qualitative questions about the patients' pain experience—"Tell me how you feel"—but ask them to rate their pain quantitatively on a Numerical Rating Scale, from 0 to 10. In contrast, ancient and medieval scientists conceived of the most relevant factors for doctors and physicists in qualitative terms: hot/cold, wet/dry. Thus, while Galen employed scales with varying degrees of hot/cold and wet/dry—where the balanced temperature was determined by combining equal amounts of ice and boiling water, while balanced viscosity was determined by combining equal amounts of ash or the like and water—he explicitly made the skin of the physician's hand, and so their personal, educated experience, the measuring tool for hot/cold and wet/dry.[4] More generally, however, ancient and medieval scientists understood hot/cold and wet/dry not as quantities measured by thermometers or on viscosity scales, but as qualitative powers to combine and to separate, or qualities of a substance to accept new configurations easily or hold on to existing ones tenaciously. An understanding of these powers was intimately tied up with the scientists' own experience as processed through their personal training and judgments.

These qualitative and quantitative approaches are two different paradigms for doing science. Indeed, at least two of the papers collected here involve scientists and sciences—astrology and dream interpretation—relying heavily upon a paradigm that appeals to the scientific role of qualities as experienced by the trained scientist. If today we prefer the quantitative paradigm for science, that is in part because perhaps we today think that quantitative concepts can increase the "objectivity" of science, whereas Islamicate philosophers did not worry that qualitative concepts would make their science "subjective," in the sense of arbitrary.

The current volume invites us to explore a world where the worries of today were not so prevalent. It invites us to challenge our notions of (1) what counts as science and (2) what counts as the experience relevant to a science, particularly the subject-rooted experience of the scientists themselves. Both today and historically, it is the scientific community that determines the answers to our two questions; however, there is no scientific community independent of the individual scientists who form it. The scientists' own skills, practices, habits, expertise, goals, and values define the norms and standards of a scientific community relative to a given science. This

[4] Galen, *De temperamentis*, 1.9, ed. Helmreich, 560–564.

relation between the individual scientists and the scientific community is reciprocal. On the one hand, the scientific community determines what phenomena or areas scientists should investigate, that is, what counts as a science, and what evidence scientists should value and use, that is, what counts as relevant experience for a science. On the other hand, the scientists themselves—their skills, practices, habits, expertise, goals, and values—define the scientific community.

There is a circle here, but it is not vicious. It is perhaps better understood as an ecosystem. Such an ecosystem ensures the objectivity of scientific experience—the individual scientists acquire their skills, practices, and habits from the scientific community, and the scientific community determines the standards by which to evaluate and judge the bona fides of individual scientists' expertise, scientific goals, and epistemic norms. Yet this scientific ecosystem is attuned to the role of the subjectivity of the individual scientists' experience, which itself forms and shapes the scientific community. This scientific ecosystem forms the basis of what the editors call "internalized objectivity," which they envision as a means for mending the divide between objectivity and subjectivity current in present-day histories of science. The hope is that internalized objectivity can provide a new lens for investigating the history of science.

Of particular interest in this picture is the place of the individual scientists and their own experiences; what the editors call the "subject-rootedness" or "subject-dependence" of experience. There is no object, event, or process that is itself simply *experience*, whatever that might mean. It is always the experience *of a subject*. Experience requires the self-aware or self-conscious (*shuʿūr bi-dhāt*) "I" that accompanies all experience, a notion that was central for Ibn Sīnā (d. 427/1037) and to the medieval Islamic scientific traditions after him.[5] The "I" for Ibn Sīnā is both the subject of experience and the complex integration of all of one's experiences, whether in the form of sensations (internal and external), memories, knowledge, or aims. Ibn Sīnā's "I" is a nexus (*ribāṭ*) of one's experiences.[6] For the practitioner of a given scientific field, then, this "I," subject, or nexus is never a pure and unencumbered "I." The "I," which is the scientist, depends upon and is the product of countless other experiences, in the form of encounters with nature, social communities, and relevant experts.

Indeed, Ibn Sīnā's "I" emerges from taking to heart Aristotle's claim at the beginning of his *Posterior Analytics* that all intellectual (Gk. *dianoētikos*, Ar. *dhihnī*) teaching and learning comes to be through pre-existing knowledge (Gk. *ex proüparchousēs gnōseōs*, Ar. *min maʿrifa mutaqaddima* or *bi-ʿilm qad sabaqa*). For al-Fārābī (d. ca. 339/950), Ibn Sīnā, and many after them, this Aristotelian dictum is seen in the fact that all intellectual instruction, whether from the experience of a teacher or self-instruction from personal experience, requires prior experience in the form of conceptualization (*taṣawwur*) and assent or verification (*taṣdīq*). For these thinkers, this prior conceptualization and assent initially come from being part of a community—first one's social and linguistic communities, to which I return shortly, and ultimately the epistemic cultures of the scientist.

[5] See Kaukua, *Self-Awareness in Islamic Philosophy*.

[6] Ibn Sīnā, *Avicenna's De anima*, ed. Rahman, 253.

By "epistemic cultures" I mean, loosely, the diverse (even area-specific) arrangement and mechanisms—whether through affinity, necessity, or historical accident—that make for "how we know what we know."[7] One important such mechanism and arrangement by which thinkers in the medieval Islamic world came to know what they knew was the exegetical commentary (*sharḥ*) genre.[8] Today we tend to view commentary as in some sense a passive exercise: the commentator lets the original author speak to them and the commentator's duty is to listen and relate the original author's intention, regardless of whether that view is true or scientifically adequate. In contrast, for ancient and medieval commentators, theirs was an active engagement with the text. The presumption was that a given authority—whether Aristotle in the Graeco-Arabic philosophical tradition of *falsafa*, Ibn Sīnā in the post-Avicennan *ḥikma* tradition, or the Qur'an and *sunna* of the Prophet in the Islamic *kalām* tradition—had hit upon the truth. Thus, it was the commentator's duty to make sure that their commentary presented the truth and was scientifically adequate regardless of the authority's original intent, although, to be sure, making it appear as if their interpretation is in fact the authority's original intent.

Ibn Sīnā—most of whose works, while not strictly commentaries, lie squarely within the tradition of commenting upon and explaining Aristotle—is a good example. In his monumental philosophical *summa* the *Shifā'* (The Healing), Ibn Sīnā at times engages in strenuous mental gymnastics and aggressively massages Aristotle's text to make the Master of those who know come out "right." This effort sometimes takes the shape of reinterpreting Aristotle's arguments in very non-Aristotelian ways or *in extremis* admitting his own inability (a rare thing for the self-assured Ibn Sīnā) to grasp Aristotle's argument properly, which he suggests must be correct but beyond his ken. For Ibn Sīnā and most intellectuals in the medieval Islamicate world, earlier commentaries were both training grounds and sparring partners for later commentaries. The commentators formed a community of like-minded thinkers who created a body of knowledge, the commentary tradition itself, that incorporated and corrected other commentaries, but always with an eye to the commentator's own subject-rooted experience regarding the true and scientifically adequate view of the world.

In addition to the community of commentators, one's linguistic community, with the conventions unique to specific languages and their role in shaping one's experience, was also recognized as significant. In fact, both al-Fārābī and Ibn Sīnā were acutely aware of the significance of the linguistic community and of engaging in discourse (*mukhāṭaba, kalām*) to acquire knowledge, whether through conceptualization or verification/confirmation.[9] Undoubtedly, their concern for the role of

[7] Knorr Cetina, *Epistemic Cultures*, 1. The notion of "epistemic cultures" is clearly indebted to the work of Karin Knorr Cetina, though I have unmoored it from its context of high-energy physics and molecular biology labs.

[8] This genre should be contrasted with the aporetic commentary (*shukūk*) genre, which raises doubts, but with the same eye to ultimately getting at the truth; for a discussion of these two genres, see Ayman Shihadeh, *Doubts on Avicenna*, 44–49.

[9] Al-Fārābī makes this point explicitly in his *Kitāb al-Burhān*, 78–79; see also Black, "Al-Fārābī on Meno's Paradox," esp. 23–28. Ibn Sīnā makes the point implicitly in his own discussion of

the scientist's linguistic community was the result of the celebrated earlier debate between the grammarian Abū Saʿīd al-Ḥasan al-Sīrāfī (d. 368 /979) and the titular head of the Baghdad Peripatetics, Abū Bishr Mattā ibn Yūnus al-Qunnāʾī (d. 328/ 940), a teacher of al-Fārābī. The debate was on the relative merits of Aristotelian logic (*manṭiq*) as opposed to the grammar (*naḥw*) of a particular language. Both participants accepted that there is a distinction between the mental sense (*maʿnā*) of a thing (e.g., snow), which may be (roughly) common among humans regardless of their language, and the corresponding verbal expression (*lafẓ*), which is unique to a language (e.g., *snow*, *neige*, *Schnee*, or *thalj*). The dispute was over whether there is a "universal grammar" of thought—in this case, Aristotelian logic—that is free of any particular language.

On a casual reading of the debate, which the belletrist Abū Ḥayān al-Tawḥīdī (d. 414/1023) reports on the basis of eye-witnesses including al-Sīrāfī, the whole exchange seems little more than a series of *ad hominem* and red-herring arguments by al-Sīrāfī to show Mattā's ignorance of the technicalities of Arabic grammar.[10] In fact, Mattā freely admits his ignorance of the finer points of Arabic grammar, but claims that all Aristotelian logic requires is a knowledge of the Arabic noun/term (*ism*), verb (*fiʿl*), and particle (*ḥarf*).[11] Al-Sīrāfī objects that no two languages exactly correspond (*lā tuṭābiq*) in how they delimit or define (*ḥudūd*) the description of terms, verbs, and particles, and such delimitation occurs only through a thorough knowledge of a specific language's syntax and semantics.[12] In effect, for al-Sīrāfī, a language determines how speakers in a linguistic community classify and categorize their experiences, and yet classification and categorization are at the heart of Aristotelian logic.

One of al-Sīrāfī's purportedly *ad hominem* grammar questions shows his point in more detail.[13] He asks Mattā whether the two sentences "Zayd is the noblest of the brothers" and "Zayd is the noblest of his brothers" are both well-formed. Mattā says they are. Al-Sīrāfī then points out that the latter, "Zayd is the noblest of his brothers," in fact is not well-formed, for while Zayd is one of *the* brothers, he is not himself one of *his* brothers; brotherhood is not a reflexive relation. Thus, to say that Zayd is the noblest of his brothers is tantamount to saying that Zayd belongs to a class to which he does not belong. Mattā made a category mistake. What al-Sīrāfī explicitly emphasizes is that Mattā's lack of knowledge of Arabic grammar has prevented him

---

how *taṣawwur* and *taṣdīq* are prior to learning and teaching in his own *Kitāb al-Burhān*, when he speaks of the prior need to conceptualize a verbal expression (*lafẓ*), term (*ism*), or statement (*qawl*) before learning can occur; see Ibn Sīnā, *Kitab al-Burhān*, ed. ʿAfīfī, 1.3, 57–58 and 1.5, 68–69; McGinnis, "Demonstrating Experience." Ibn Sīnā is more explicit about the place and role of the linguistic community and the mutual agreement or convention (*tawāṭuʾ*) surrounding terms in his *Kitāb ʿIbāra*, 1.1–2.

10 The Arabic and an English translation (the latter is modified here) are available in Margoliouth, "The Discussion."

11 Margoliouth, "The Discussion," 98 (Ar.)/117 (En.).

12 Margoliouth, "The Discussion," 99 (Ar.)/118 (En.).

13 Margoliouth, "The Discussion," 103 (Ar.)/121–22 (En.).

from correctly identifying the proper genus (*jins*).[14] Here it is important to note that al-Sīrāfī is not using "genus" in a non-technical way, but intends the logical notion of genus, which is central to Aristotelian logic.

The centrality of the genus (the race, stock, or family—loosely, a class of things) to Aristotelian logic in al-Sīrāfī's example deserves further elaboration. Aristotle's logic is a categorical logic. It categorizes our experiences into classes of existents and then considers the relations among these classes. At its root, categorical logic tells one what forms of inference are valid, that is, are truth-preserving, among the various relations in which classes can stand to one another, *if* one starts with a true classification of one's experience. Categorical logic is silent, however, as to the true classification of our experience. Al-Sīrāfī's point is that not only can a proper grounding in grammar guide one in determining valid inferences, but it also helps one to start with a proper classification of experience. Thus, knowledge of a language-specific grammar is superior to knowledge of Aristotelian logic.

Al-Sīrāfī's point throughout the debate is that Aristotelian logic is all but useless if it cannot help one categorize, classify, and interpret one's experiences in ways that others can understand. Providing this help is precisely what Aristotelian logic does not do, complains al-Sīrāfī, whereas language-specific syntax and semantics does do it. Thus, al-Sīrāfī chides Mattā:

> When you [Mattā] say to someone, "Be a logician," you mean be rational [*ʿaqlī*] or a reasoner [*ʿāqil*] or rationally understand [*iʿqil*] what you are saying [...]. When another says to you, "Be a grammarian, a philologist, and use the language correctly," he means understand what you are saying yourself and then seek to make others understand you, making the verbal expression [*lafẓ*] measure up to the mental sense [*maʿnā*] such that the former does not fall short of it.[15]

Here we see that al-Sīrāfī vividly recognizes the internalized objectivity of experience: the subject-rooted experience of the individuals shapes and forms a linguistic community, but the linguistic conventions and standards of that community determine proper classification and the practices of the individuals.

Al-Sīrāfī's point did not fall on deaf ears, as Ibn Sīnā, writing a generation or more after the al-Sīrāfī–Mattā debate, gives witness. While Ibn Sīnā defends the universal nature of Greek logic as a tool for reasoning and weighing the validity of arguments, he also recognizes the place of specific languages and their conventions in scientific discourse, however construed. A few examples from his natural philosophy and logic make my point.

One is the passage where Ibn Sīnā comments on Aristotle's discussion in *Physics* 1.7 concerning one thing's coming to be *from* (Gk. *ek*, Ar. *ʿan*) another. Ibn Sīnā provides a detailed account of the nuances of the Arabic prepositions *ʿan*, *baʿda*, and *min* and their adequacy and inadequacy in explaining Aristotle's point. In the end, he submits, "Still, I do not insist on this and similar cases, since languages may differ in the license and proscription of these uses."[16] Another example, and

[14] Margoliouth, "The Discussion," 103 (Ar.)/122 (En.).

[15] Margoliouth, "The Discussion," 106 (Ar.)/125 (En.).

[16] Ibn Sīnā, *The Physics of The Healing*, ed. McGinnis, 1.2 [para. 20].

arguably Ibn Sīnā's most significant acknowledgment of the role of language in interpreting our experiences, can be found in his *Kitab al-Burhān* (Book of Demonstration).[17] There Ibn Sīnā combines Aristotle's scientific questions or objects of inquiry (Gk. *zētoumena*, Ar. *maṭālib*) from *Posterior Analytics* B.1—the *what*-question, *that*- (and *whether*-) question(s), and the *why*-question—with the distinction between the two basic types of knowledge, *taṣawwur* (conceptualization) and *taṣdīq* (assent, confirmation, or verification). Prior to a scientific investigation into causes (that is, the *why*-question), Ibn Sīnā insists that one must answer the *what*-question, which involves conceptualization, and the *that*-question, which involves assent or verification. Moreover, for Ibn Sīnā, there are two levels of *what*- and *that*-questions. The first level is what might be considered the proto-scientific answers to the *what*- and *that*-questions, namely, answers that draw upon the subject-rooted experiences of those within a particular *linguistic community*. The second level is the proper scientific answers based on the internalized objectivity of one's *scientific community*, which shapes and constrains one's answer.

For figures like Abū Maʿshar, al-Fārābī, Ibn Sīnā, and others, science ultimately seeks causes: answers to *why*-questions. Such a search, however, presupposes the prior knowledge that there is a causal link between events, which itself requires conceptualizing and then confirming the essential definitions of the phenomena involved in any given causal nexus, a process informed and partially constructed by a community of experts. Acquiring the essential definitions, in its turn, presupposes conceptualizing and then verifying that something in fact corresponds with those phenomena. This foundational stage of conceptualization and verification is rooted both in the conventions of the linguistic community, which provides the initial categorization and classification of one's experiences, and in one's own subject-rooted experience, which appeals to the internalized objectivity that the community of experts shapes and confirms.

The above discussion was intended to provide a very general introduction and one specific way that medieval Islamicate thinkers' own thoughts resonate with the editors' observation about subject-rootedness and the internalized objectivity as it played out in medieval Islamic science. Let me now turn to the other three papers that make up this volume and consider how those same themes appear in other scientific contexts.

Charles Burnett's "'Obvious, Clear, and in Front of Our Eyes'" rehearses Abū Maʿshar's appeal to experience in his defense of astrology. As Burnett notes (p. 22), Abū Maʿshar's introductory book to his *Great Introduction to Astrology* is modeled on Ptolemy's *Tetrabiblos*, even if the *Great Introduction* goes into considerably more detail, and so broadly falls within the community of exegetical commentaries. Ptolemy already recognized that the possibility (Gk. *dunaton* $\approx$ Ar. *quwwa*) of astrology as a science was in question. Though Ptolemy offers a defense of astrology's scientific standing, Abū Maʿshar must have felt it was insufficient, and so offers a

[17] The following discussion is drawn primarily from Ibn Sīnā, *Kitāb al-Burhān*, 1.3 and 1.5–1.6.

greatly extended response to the detractors of astrology. In fact, Abū Maʿshar catalogues ten different kinds of rejection, ranging from the outright denial that celestial objects can provide any indications or signs concerning terrestrial phenomena, to laments concerning the prevalence of charlatans and dilettantes, to more sophisticated complaints about the discrepancies among the handbooks of astronomical tables (Zījes) used to tabulate the parameters needed for astrological predictions or the practical impossibility of acquiring sufficient experiential evidence to confirm astrology as a science.[18] In short, Abū Maʿshar needed to answer the *that*-question: whether there is a genuine science of astrology rather than a mere pseudoscience.

Abū Maʿshar explicitly acknowledges and intentionally follows Aristotle's four objects of scientific inquiry.[19] Moreover, he draws upon experience (*tajriba*) much as Ibn Sīnā will do in his discussion of veridical dreams. Abū Maʿshar first appeals to one's own experience—both the subject-rooted experience of the individual and the internalized objectivity that a community of experts affords—to validate his subject matter, then proceeds to the scientific investigation of the heavenly bodies and only afterwards to their influence on terrestrial affairs.

Astrology, Abū Maʿshar tells us, is the science of judgments (*aḥkām*) concerning past, present, and future events based upon the signs and indications observed from the motions and impressions of the powers of celestial bodies on terrestrial objects.[20] One's own experience, Abū Maʿshar believes, confirms that celestial bodies are active and influential in terrestrial affairs. He provides two categories of analogical arguments (*qiyāsāt*) drawn from experience to validate the fact-*that*. The two categories appeal to experiences that are (1) obvious to the common person (*ʿāmma*) and (2) not so obvious, but which experts even outside astrology grasp. These arguments roughly appeal to subject-rooted experience and to the internalized objectivity supplied by the expert testimony of a broadly scientific community.

Abū Maʿshar's arguments from the first category, which appeal to the subject-rooted experiences of all, include the obvious fact that the motion of the sun affects the qualities of hot, cold, wet, and dry here on Earth, whether daily or over the seasons.[21] Moreover, one clearly recognizes that these effects on those qualities in turn affect animals, plants, and even minerals. Similarly, many recognize the effect of the moon and its cycles on climatic conditions, menses, the rising of the seas, and the like, all of which affect animals and plants. The sun and moon are the two great celestial wanderers (Gk. *planētes*). Given the obvious effects of these two celestial bodies on terrestrial conditions and events, it would simply be obtuse, maintains Abū Maʿshar, to deny similar powers to the remaining planets, for it is the motions of the

---

[18] Abū Maʿshar, *The Great Introduction*, 1.5, 107–149.

[19] Abū Maʿshar, *The Great Introduction*, 1.3, [3.2b], 81; bracketed numbers refer to the marginal enumeration of Yamamoto and Burnett's edition. While both Abū Maʿshar and Ibn Sīnā share the same list of objects of scientific inquiry, they identify those four questions slightly differently. According to Abū Maʿshar, the four are the *what* (*mā*)-question, the *whether* (*hal*)-question, the *how* (*kayfa*)-question, and finally the *why* (*lima*)-question.

[20] Abū Maʿshar, *The Great Introduction*, 1.2 [2.4], 55.

[21] Abū Maʿshar, *The Great Introduction*, 1.2 [2.6a–2.15], 55–61.

other planets that explain the nuanced daily and yearly differences that we experience in weather and the like.

His arguments from the second category, which appeal to the testimony and internalized objectivity furnished by non-astrological experts, include those of farmers, herdsmen, sailors, and others.[22] These experts have subtle experiences, based upon their trained observations of planetary changes, that allow them to know the best time to sow, to breed livestock, or to sail. All of these experts validate the fact-*that* the heavens and their motions affect terrestrial events.

The expert most like the astrologer is the doctor, and yet, claims Abū Maʿshar, astrology is nobler than medicine.[23] He reasons thus: The doctor uses signs and indications observed in the patient to predict their state of health or sickness. Health and sickness, according to the humoral medicine of the time, are determined by reference to whether, relative to the individual patients, they are in a well-balanced humoral state (health) or have an excess or deficiency with respect to some humoral state (sickness). The four basic humors of ancient and medieval medicine are themselves determined by their hot–cold, dry–wet natures: blood is a hot/wet mixture; phlegm is a cold/wet one; yellow bile is a hot/dry mixture; and black bile is a cold/dry one.

Later in the *Introduction*, Abū Maʿshar draws a distinction, which is useful here, between an *effect* (*mafʿūl*) of another and *being acted on*, *influenced by*, or *affected by* (*yanfaʿil ʿan*) another.[24] An effect is the end result of the agent's action on a patient, such as a built wall or fire's burning, and, while Abū Maʿshar does not say as much, effects seem to be necessitated by their causes, whether the active power of the agent or the passive power of the patient. In contrast, being acted on or affected is the response of the patient (that is, the one being acted upon) to the agent's action. The patient's response is very much conditioned by the patient's state, their elemental mixture or humoral temperament. In other words, the effect is the product of *both* the agent's action and the patient's state. Thus, volitional patients can act in ways that would alter their state, and so perhaps thwart, mitigate, or in some way alter the effect of the agent, were the patient not to act otherwise. The fact that there is a difference between an effect and being acted on is what gives medicine its practical value. From the signs and indications that the doctor observes, a course of treatment can be suggested that may cure or lessen the ailment or, in the worst case, give the patient the opportunity to put their affairs in order.

The astrologer, on the other hand, uses signs and indications observed in the heavenly motions to predict the state of the hot/cold and wet/dry conditions not only of the climate but potentially of any natural terrestrial object. That is because, according to the natural philosophy of the time, all mixtures, like humors, are composed of four elements, whose nature is again understood in terms of the qualities hot/cold and wet/dry: elemental fire (hot/dry), air (hot/wet), water (cold/wet), and earth (cold/dry). Moreover, the heat and coolness as well as the wetness and dryness of the ambient environment affect these elements so as to bring about new mixtures and humoral

---

[22] Abū Maʿshar, *The Great Introduction*, 1.2 [2.12–2.17], 59–63.

[23] Abū Maʿshar, *The Great Introduction*, 1.2 [2.23c–2.26e], 67–73.

[24] Abū Maʿshar, *The Great Introduction*, 1.3 [3.7–3.9], 87–91.

states. Just like that of the doctor, the advice of the astrologer, based upon signs and indications, can aid the client to act in a way that can alter how they are acted on by the celestial powers, whether to avoid or mitigate harm or at least prepare for the worst. The significant difference between the astrologer and the doctor is that the astrologer considers the signs and indications of the *causes* of these primary qualitative states on Earth, where again these causes are the motion of the heavenly bodies, whereas the doctor considers the signs and indications of the *effects* of these primary qualitative states, namely, health and sickness.

To bring the discussion back to the testimonial experience of the *that* concerning astrology, no one doubts that there is a science of medicine that has as its subject matter a certain effect. Thus, no one should doubt *that* there must be a nobler science that has as its subject matter the cause of that effect, and this science is astrology. Thus, one's own experience with medicine, coupled with a proper understanding of its value, answers the *that*-question concerning the possibility of a science of astrology.

Another doubt about whether there can be a science of astrology is particularly interesting because it strikes at precisely the use of experience (*tajriba*) to confirm astrology's credentials.[25] The doubter observes that one can appeal to experience only if multiple experiences make up one's evidential basis. Yet it takes thousands of years for the configuration of the planets and stars at a given time to return to a similar configuration in the future. Consequently, no one would ever have the multiple experiences required to confirm astrology. Abū Maʿshar's response is to appeal not to the personal experience of a single astrologer, but to the collective internalized objectivity of the community of astrologers and their testimony over the course of centuries and even a millennium. The community of astrologers can discover general trends over long periods and through analogical reasoning (*qiyās*) applies them to particular cases. Abū Maʿshar concludes his response, "Truthful reports about a given thing's occurring again and again at different times and clear indications stand in for observation and the immediately present."[26] Abū Maʿshar's point confirms the observation of the editors of the current volume: experience is for Abū Maʿshar subject-rooted, and the internalized objectivity provided by a community of experts is what shapes and forms that subject-rooted experience.

Jules Janssens's contribution, "Dream-Experience in Ibn Sīnā," takes up the common experience of dreams and Ibn Sīnā's analysis of veridical dreams in particular. In many respects, Ibn Sīnā's defense of veridical dreams parallels Abū Maʿshar's defense of astrology. Both thinkers feel obligated to confirm the veracity of their respective subject matters. Both assert that in fact, once properly understood, common experience verifies and confirms the truth claims of their respective subjects. Both appeal to a similar view of science, particularly that grounded in Aristotle's *Posterior Analytics* and its account of the proper objects of scientific inquiry. Both thinkers attempt to ground their subjects in the best natural philosophy of their time,

---

[25] Abū Maʿshar, *The Great Introduction*, 1.5 [5.25–5.31], 129–133.

[26] Abū Maʿshar, *The Great Introduction*, 1.5 [5.31], 133.

namely, a particular theory of the elements and humoral medical theory and those sciences' accompanying categories. Let me briefly comment on these parallels in turn.

The notion of prophetic dreams, interpretative dreams—that is, dreams that explain the meaning of an earlier dream—and even "simple" veridical dreams might sound uncommon or even paranormal to some. Ibn Sīnā assures us that we have good experiential evidence for their existence. To be sure, prophetic and interpretative dreams are quite likely rare, certainly in the case of genuinely prophetic visions. Still, what I have called simple veridical dreams are, Ibn Sīnā tells us, more common than one might unreflectively think. In fact, Ibn Sīnā contends that there is no one who has not had their share of such dreams.[27] In his autobiography, he provides a personal example that might resonate with others: He notes that when he was grappling with a particularly difficult philosophical problem, he would put his whole mind to it during his waking hours. Then, "whenever sleep seized me, I would see those very same problems in my dream, and many of those problems became clear to me in my sleep."[28]

The experience is common enough. As Ibn Sīnā observes, when our waking hours are thoroughly occupied with some puzzle or topic, we continue to try and work through it in our sleep. The phenomenon Ibn Sīnā describes is seen in the "*Tetris* effect," when people experience the various blocks, or tetrominoes, of the game they played throughout the day falling in their dreams and even continue the game's puzzle-solving activity in their dreams. (Interestingly, this effect can occur even in a wakeful state, a point that Ibn Sīnā himself recognizes in his analysis of dreams.) There are also richer examples of (modern) veridical dream-experiences. One need only mention August Kekulé (1829–1896) and his dream of a snake's eating its tail (the ouroboros), which purportedly led to his uncovering the ring structure of benzene. Here some truth about reality was anticipated in Kekulé's dream-image.

My choice of examples is intentional. I was subtly (or not so subtly) trying to get the reader to recognize *that* there is a certain phenomenon worthy of scientific inquiry, namely *that* there is, or at least can be, a relation between certain mental experiences, dreams (and waking visions), and extra-mental experiences, the way things are in the world. My examples were drawn from testimony (*tasāmuʿ*), as in Kekulé's dream, and the personal experience (*taʿāruf*) of many, as in the *Tetris* effect. Ibn Sīnā identifies both the testimony of experts and subject-rooted experience as the experiential evidence that underlies his notion of *tajriba* or systematic or methodic experience, and it is precisely to *tajriba* that Ibn Sīnā appeals to verify or confirm the existence of veridical dreams.[29]

Tajriba, for Ibn Sīnā, helps to answer the scientific *that*-question and to establish *that* some causal link exists between two phenomena, even if it does not uncover what that link is. Ibn Sīnā had need of *tajriba* to verify and confirm the experience of veridical dreams, for it is simply naïve to think that people of Ibn Sīnā's time

[27] Ibn Sīnā, *Avicenna's De anima*, 4.2, ed. Rahman, 174.1.

[28] Ibn Sīnā, *The Life of Ibn Sīnā*, ed. Gohlman, 30–31; translation modified.

[29] Ibn Sīnā, *Ishārāt*, ed. Forget, *namaṭ* 10.8.

were any more credulous than people today. Certainly, there were those who would blindly accept the reality of veridical dreams, but so are there those today who would blindly accept them (along with crystal healing and microchips in COVID vaccines). Still, many do not, and the same was true in Ibn Sīnā's time. In fact, *namaṭ* 10 of the *Ishārāt wa-tanbīhāt* (Pointers and Reminders), to which I return shortly, is directed precisely at those who would deny veridical dreams. Thus, Ibn Sīnā has recourse to experiential evidence in the form of testimony or internalized objectivity that the community provides and one's own subject-rooted experience precisely to establish *that* there are veridical dreams and the certainty (*yaqīn*) that some causal relation exists between dream states and the external world.

Before turning to *namaṭ* 10 of Ibn Sīnā's *Ishārāt*, an aside into Ibn Sīnā's canonical account of *tajriba* is warranted to understand exactly what sort of knowledge he thinks *tajriba* supplies.[30] That account comes at *Kitāb al-Burhān* 1.9, where he contrasts *tajriba* with induction (*istiqrāʾ*).[31] Understanding the difference between the two is useful. According to Ibn Sīnā, induction purports 1) to answer the *what*-question, namely, to identify the essential definition of a thing, and 2) to establish an absolutely (*muṭlaqan*) necessary relation between the subject and predicate terms of the essential definition derived through induction. *Tajriba* is humbler in its aspirations. It merely attempts (1′) to answer the *that*-question, namely, to indicate that certain accidental features of, or occurrences among, things are regularly related to one another, in which case one can infer the necessity of some cause for that regularity, and (2′) to show that this regular relation is conditionally (*bi-sharṭ*) necessary. Here the sensible accidents in question are not coincidental features, such as a certain color, shape, or texture, but necessary concomitants (*lawāzim*) following from the thing's nature, such as fire's heating or alcohol's intoxicating. The conditions for the conditional certainty that *tajriba* affords include geographical locations, the state of the elemental or humoral mixture, the ambient environment and its effect on what is being observed, and so on. These are the same conditions that Abū Maʿshar had identified as conditioning judgments about the stars' and planets' power to affect or act upon us. What is important to stress is that for Ibn Sīnā, *tajriba* addresses neither the *what*-question nor the *why*-question. In other words, it neither provides an essential definition nor reveals the necessary causal link—the middle term—that explains the observed regularities. This cause or middle term is discovered only through syllogistic reasoning (*qiyās*).

Ibn Sīnā's discussion of veridical dreams in *namaṭ* 10.7–8 of the *Ishārāt*, though brief, shows the whole scientific process outlined here in actual practice. While Janssens has translated the most relevant passage in part, it does not hurt to repeat it here. I present it in its entirety and with the immediately preceding *ishāra* or "pointer" to provide context:

---

30 Studies of *tajriba* in Ibn Sīnā include McGinnis, "Scientific Methodologies," and Janssens, "'Experience' (*tajriba*)."

31 Ibn Sīnā, *Kitāb al-Burhān*, 1.9, ed. ʿAfīfī, 95–98; English translation in McGinnis and Reisman, *Classical Arabic Philosophy*, 147–152.

> *Pointer* [1]: When you hear that a master [*ʿārif*] reports a hidden matter [*ghayb*] and then the good news or advance warning turns out true, accept it and do not be dubious, for there are known causes for that among the schools of natural philosophy.
>
> *Pointer* [2]: Experience and syllogistic reasoning [*al-tajriba wa'l-qiyās*] are in mutual agreement that the human soul can acquire something of hidden matters while dreaming. (There is no obstacle that one cannot circumvent or surmount to acquiring something like that while awake.) As for experience, testimony and personal acquaintance [*al-tasāmuʿ wa'l-taʿāruf*] both give witness to it. Everyone has experienced that himself through multiple experiences, which inspire him to assent to [the reality of veridical dreams]. That is, unless he is one of those with a corrupt temperament, whose faculties of the compositive imagination and memory are dormant. As for syllogistic reasoning, reflect on it among the [subsequent] reminders [*tanbīhāt*].[32]

Pointer 1 indicates that there were those who found the reality of veridical dreams hard to believe. Ibn Sīnā assures them that they should accept the existence of and reports about such dreams, for these can be fully explained by appeal to natural causes. In pointer 2, Ibn Sīnā appeals to testimony, namely that of the community of experts, and personal acquaintance to answer the *that*-question concerning veridical dreams. There is a necessary link between dream states (or sometimes waking visions) and certain truths about the world yet unknown. Indeed, the link is sufficiently necessary that, he claims, everyone can have this experience. Second, he names the conditions under which one would *not* have such an experience, or at least when dreams might not appear as veridical. Those conditions present obstacles to such an experience, which Ibn Sīnā identifies with cases of humoral imbalance where the compositive imagination and memory are inactive or sluggish. Finally, he indicates that he will remind the reader of the salient premises needed for a demonstration-*why* that reveals the necessary links between the dream state and reality.

The subsequent reminders provide the general account of dreams that unfolds in Janssens's discussion, which can be quickly summarized here. Ibn Sīnā's account has two steps. First, during sleep the human soul encounters celestial souls, which produce similitudes or traces of the causes of events on Earth in the human soul. (While the view that higher-order human cognition requires contact with a separate substance or celestial soul might sound paranormal to us today, it was part and parcel of the communities both of natural philosophers and of Aristotelian commentators going back at least as far as Alexander of Aphrodisias [fl. ca. 200 CE] and Themistius [d. 387 CE], to which al-Fārābī, Ibn Sīnā, and others in the *falsafa* tradition were heir.) Second, the imagination cloaks the intelligible forms with sensible forms or reinterprets them into sensible forms. Such an account certainly allows for veridical dreams because dreams, on this account, really are the sensible representations of truths about our world. However, if the dreamer was not focused on a particular problem during waking hours, what is seen while in contact with the celestial realm itself will not be focused and may come from any of the issues that crossed the dreamer's mind prior to sleep. Moreover, since the compositive imagination is localized in the brain, the temperamental or humoral balance of the individual affects the images that cloak what is seen while dreaming. The temperamental or humoral

32 Ibn Sīnā, *Ishārāt*, *namaṭ* 10.7–8, 209–10.

balance is itself affected by one's natural temperamental state, the ambient environment, diet, and the like. Both whether the dream has been focused on a particular intellectual activity and one's temperamental or humoral state affect the images of the dream and so can lead to "muddled dreams."

Let me conclude this section by showing how the preceding model of veridical dreams additionally explains interpretative dreams and prophetic visions. In the case of interpretative dreams, the object of focused intellectual activity during the waking hours is the correct interpretation of some initial dream. Subsequently, this activity of intellectual interpretation continues when one sleeps and so directs what one sees in the celestial realms, which again is re-cloaked in images. The result is a second dream about the meaning of the first dream. As for prophetic visions, the individual may have spent a day in deep meditation on, for example, what is required to be in right relation with God, which again directs what they see when in contact with the celestial soul. Since for Ibn Sīnā a necessary condition of a prophet is a well-balanced temperament or humoral mix, in the case of the prophet the images and their interpretation are less veiled, and so clearer. Thus, the intellectual content seen in the celestial realm may be cloaked merely in an angelic or humanlike image that speaks directly about the intellectual content, where the sounds heard are themselves aural images immediately recognizable by the dreamer. (When this contact and the ensuing images occur during a waking state, as again Ibn Sīnā allows, there is a prophetic vision.)

In "Translating Epistemic Norms into Social Hierarchy," Nimrod Hurvitz considers three dimensions of the *miḥna*, a period of religious interrogation lasting from 833–851 CE. The dimensions that Hurvitz identifies are three:

1) the differing epistemic norms that the historic actors held, namely, those of the proponents of *kalām* or speculative theology (*mutakallimūn*), specifically Muʿtazilī *kalām*, and those of the proponents of *ḥadīth* or Prophetic reports and traditions (*muḥaddithūn*), and the differing practices that emerged from them;
2) the hierarchical power struggle that ensued from deciding between these epistemic norms, namely, who is most suited to provide the correct understanding of Islam and its creeds; and, finally,
3) the emotional strife that this power struggle created, namely, the mutual fear, anger, and disgust that came to exist between the two sides.

In these notes, I consider primarily the first dimension, the differing epistemic norms of the two groups, and make only very brief comments about (2) and (3) at the end. Hopefully, my comments will contextualize those norms as they relate to both subject-rooted experience and internalized objectivity.

"Epistemic norms" are the standards or rules used to determine whether one's beliefs are rational, that is, whether one's holding a belief or set of beliefs is judged reasonable by others within the community. That is, epistemic norms determine what counts as good justification or warrant for a given belief or set of beliefs.[33] For

[33] For a discussion of the contemporary philosophical debate, see Pollock, "Epistemic Norms."

the *mutakallimūn*, these epistemic norms centered on the use of *qiyās*, whereas the *muḥaddithūn* relied on *taqlīd*. *Qiyās* is commonly understood in terms of argument, logical inference, and in general "reason" or being rational. In contrast, *taqlīd* is frequently understood as "blind faith" and appealing to authority as opposed to independent reasoning. So understood, the competing epistemic norms of our debate look to be between *appealing to reason*, and so being rational, and *not appealing to reason*, and so being irrational.

This dichotomy is certainly the way that the *mutakallimūn*, and even some contemporary scholars, see the debate. Setting up the dispute in these rhetorically charged terms, however, puts the *muḥaddithūn* in an awkward position, for they now seem to hold the internally contradictory claim that the standard for being rational is that one be irrational. Of course, the *muḥaddithūn* could have been proto-Kierkegaardians; I believe because it is absurd. I suggest, however, that there are other ways to understand the competing epistemic norms, which perhaps offers a more nuanced understanding of *qiyās* and *taqlīd*, seen now in terms of different kinds of experience, differing views of what counts as rational, and differing views of the role of community in rational discourse about belief.

The different kinds of experience to which the *mutakallimūn* and *muḥaddithūn* appeal, I submit, are nothing other than those already seen at work in the discussions of Abū Maʿshar and Ibn Sīnā, who, as we have seen, distinguish between subject-rooted experience and testimony understood as the internalized objectivity manifested in the community authorities. (Interestingly, while Abū Maʿshar and Ibn Sīnā saw subject-rootedness and internalized objectivity as working in tandem to form an organic whole, the participants in the later debate seem to have wanted to separate them, and this very separation, I argue, is what gave rise to the heated nature of the controversy.) For the *mutakallimūn*, subject-rooted experience underlies correct epistemic norms. This experience takes the form of one's own sensation (*ḥiss*) and natural intuition (*gharīza*) of necessary truths, then reasoning (*qiyās*) from one's sensible experiences and/or perceived necessary truths to otherwise unknown sets of beliefs. In this respect, the *mutakallimūn* are foundationalists; some beliefs are inferred from other beliefs, but there is an ultimate set of basic beliefs that are foundational and are known directly. On this view, one is rational if and only if the evidence supports one's beliefs. In other words, one's beliefs must be grounded in one's experience of the world around one, and one's experience of the world is acquired through the senses or known immediately or inferred from beliefs so grounded. The experiential evidence of the senses, natural intuition, and logic, understood as the laws of thought or correct reasoning, are, for the *mutakallimūn*, applicable to all and hold in every sphere of discourse. To forego using these God-given tools when interpreting Islam is to sink into irrationalism.

In contrast, the *muḥaddithūn* give pride of place to testimony, which has as its focal point the internalized objectivity afforded by a community of authorities, most specifically the authority of the Qur'an and the *sunna* of the Prophet Muḥammad. This reliance on sacred scripture and tradition has implications for what counts as rational. Rationality is no longer determined by some external set of criteria that, as it were, has universal validity for all and in all contexts, as the *mutakallimūn* would have it. Instead, being rational is understood as possessing an internally coherent set of beliefs. In this respect, I suggest, the *muḥaddithūn* have adopted something close to what we now consider a coherentist theory of justification.[34] The internal coherence of the set of beliefs embedded within the Qur'an and *sunna* underlies the *muḥaddithūn*'s experience of the divine and provides evidence about God. Such a view explains the words of Ibn Ḥanbal, discussed by Hurvitz (p. 61):

> The *sunna*, in our view, is the vestiges of the Messenger of God (peace be upon him). The *sunna* comments and provides guides to the Qur'an. There is no *qiyās* with respect to the *sunna* nor similar things mixed with it. Neither intellects [*ʿuqūl*] nor passions grasp it; there is only following [the *sunna*].[35]

On certain issues, these resources and authorities simply do not speak. In some cases, these silent issues may appear to undermine the internal coherence when measured against some external source, such as one's experience of how key terms in the sources refer to the created order. Concerning such issues and terms, the sources and authority are silent about the *how*-question, and so based upon those sources and authority the issue or term is "without how" (*bi-lā kayf*). For these issues and terms, the *muḥaddithūn* simply refuse to speculate, for to speculate would just be to give one more narrative in a sea of "just-so stories." Such a refusal may appear to be a retreat into fideism or even irrationalism, but it can also be seen as an expression of epistemic humility. One does not have the knowledge or authority to speak on such an issue, and so one does not. In contrast, to take such pride in one's rational abilities that one is willing to harm and persecute others, as, for example, by instituting the *miḥna*, can be seen as hubris—not merely an intellectual vice but a moral one as well.

The debate between the *mutakallimūn* and the *muḥaddithūn* so understood now looks quite similar to that between Abū Bishr Mattā and Abū Saʿīd al-Sīrāfī a hundred years later: Are there objective standards of justification that have universal validity (and even forms of reasoning that apply to God) or is all discourse grounded in the conventions, practices, and indigenous sources of a given community? The issue, then, is not so much about authority, for both parties accepted the authority of the Qur'an and *sunna* of the Prophet; rather, it is whether anything more than this authority is needed within the community of Muslim believers for rational discourse about God. I would hazard that the *muḥaddithūn* do not reject reason outright in all spheres of discourse. As Hurvitz rightly emphasizes, "even these conservative thinkers [i.e., the *muḥaddithūn* and others] did not regard scientific knowledge in and of itself as unacceptable or as a body of knowledge that ought to be removed from Muslim societies" (p. 59). The *muḥaddithūn* object only to introducing elements

[34] The Ashʿarī theologian Imām al-Ḥaramayn al-Juwaynī (d. 478/1085) may have been the first to explicitly develop or at least suggest a coherentist theory of justification and rationality within the medieval Islamic world; for now, see Siddiqui, *Law and Politics under the Abbasids*, Chap. 4. Al-Juwaynī seems to be working out trends and implications already present in the earlier medieval debates about faith and reason.

[35] Ibn Abī Yaʿlā, *Ṭabaqāt*, ed. ʿUmar, 1: 337; here I have provided my own translation and some more of the text.

foreign to the Qur'an and *sunna*, whether a universal logic or Greek science and philosophy or even one's own speculation, into the theological discourse specific to the Islamic community.

The *muḥaddithūn*'s limits on the spheres of theological speculation and their reliance solely on authority (*taqlīd*) in these matters are not obviously arbitrary nor a matter of blind faith. At the heart of speculation (*naẓar*) for the *mutakallimūn* is *qiyās*. The root meaning of *q-y-s* is "to measure," and so *qiyās* implies a measure or standard. Indeed, when Aristotle's logical works were translated into Arabic, *qiyās* was used for the Greek *sullogismos*, that is, the syllogism or pattern of inference that is the standard by which valid thought is measured. The notion of an Aristotelian syllogism, however, is not how the *mutakallimūn* nor the *muḥaddithūn* of our debate would have understood *qiyās*. Instead, they would have understood it much in the way that Abū Maʿshar employed it in his defense of astrology, namely, as a form of analogical reasoning. (Indeed, Abū Maʿshar was a *ḥadīth* scholar in Baghdad, the seat of Abbasid rule, during the time of the *miḥna*.)

Qiyās, in fact, is a term of art in Islamic jurisprudence (*fiqh*), which the *mutakallimūn* extended to theology.[36] While *qiyās* subsumed various forms of reasoning, the unquestionably most important of these is again analogical reasoning. At the heart of analogical reasoning in its technical, legal sense is the notion of the *ʿilla* or explanatory reason (in law, the *ratio legis*) that exists between the two cases and allows one to construct the analogy. For any argument by analogy to be cogent, there must exist some core, common similarity (*ʿilla*) between the two cases being considered, otherwise the analogy is weak and so commits an informal fallacy.

Therein lies the problem. In reference to God, the Qur'an states, "There is nothing like Him" (*laysa ka-mithlihi shay'un*) (Q. 42:11). This verse and its implications are central to much of the theological discourse during the formative years of Islamic theology, particularly with respect to God's transcendence (*tanzīh*). This Qur'anic source simply claims that nothing in the created order can be compared to God. If the verse is taken literally, and the *muḥaddithūn* certainly would have done so, then there can never be a core, common similarity (*ʿilla*) between God and anything in the created order. Consequently, any attempt to extend one's understanding of God based upon sensible experiences drawn from creation and inferences starting from those experiences is doomed to failure. All such reasoning ultimately commits the fallacy of weak analogy. Thus, to base one's theology on such reasoning, once it is viewed as fallacious, is the apex of irrationality, or so the *muḥaddithūn* could argue.

There are additional consequences of taking God to be unlike anything in creation, which perhaps explain both why the issue of the createdness of the Qur'an was central to the debate between the *muḥaddithūn* and *mutakallimūn* and why that debate was so emotionally charged. I have suggested that at the core of the debate between these two groups is a difference in the kinds of experiences that one privileged: subject-rooted experience—experienced through one's own intellectual striving and judged against a universal standard of rationality—and testimonial experience—experience shaped

[36] For an extended discussion of *qiyās* in Islamic jurisprudence, see Hallaq, *Islamic Legal Theories*, 83–95.

by the skills, practices, habits, and expertise embedded within the authoritative claims of the Qur'an and reports of the Prophet. If the Qur'an is created and so not the literal speech (*kalām*) of God, and yet nothing in the created order is like God, then the words of the Qur'an cannot be like or meaningfully refer to God. They can reveal nothing about God, what God is like, what God commands of us, or anything else, for the words of the Qur'an, if created, would be wholly unlike God. Alternatively, one can accept that the words are the literal speech of God, and if they are in need of special interpretation, one has an authoritative interpreter in the Prophet. In short, if the Qur'an is created, then for the *muḥaddithūn*, its claims—to say nothing of the claims of the *sunna* of the Prophet—would lack authority, for they would be based on weak analogies. In such a case, one simply could have no authentic testimonial experience about God, or so it would appear to the *muḥaddithūn*.

In the eyes of Ibn Ḥanbal and other *muḥaddithūn*, the *mutakallimūn*'s doctrine of the createdness of the Qur'an is the ultimate gaslighting. It denies the very basis of their experience with God. Conversely, the *muḥaddithūn*'s rejection of human reason and one's intellectual power to strive to experience and understand God denies the very basis of the *mutakallimūn*'s experience with God. It is no wonder that, as Hurvitz observes, emotions ran high. Each group viewed the other as invalidating their experience. Similarly, the hierarchical struggle that Hurvitz notes, seen in this light, need not be solely about political power (although surely there were those for whom this was the issue), but about validating one's experience of God.

I end by simply noting that this epilogue recounts my experience with the three papers and introduction that make up this volume and their treatment of experience in the medieval Islamic worlds. Others may have different experiences. Therein lies the value: experiencing experiences both rooted in the subject and shaped by a community of scholars.

Bibliography

Primary Sources

Abu Ma'shar. 2019. *The Great Introduction to Astrology*, edited and translated by Keiji Yamamoto and Charles Burnett. 2 vols. Leiden: Brill.

al-Fārābī. 1987. *Kitāb al-Burhān*n, ed. Majid Fakhry. Beirut: Dār al-Mashriq.

Galen. 1904. *De temperamentis* (*Peri kraseōn*), ed. Georg Helmreich. Leipzig: Teubner.

Ibn Abī Yaʿlā. 1998. *Ṭabaqāt al-fuqahāʾ al-Ḥanābila*, ed. Alī Muḥammad Umar. 2 vols. Cairo: Maktabat al-Thaqāfa al-Dīniyya.

Ibn Sīnā. 1956. *Al-Shifāʾ, al-Manṭiq, Kitāb al-Burhān*, ed. A. Afīfī. Cairo: al-Maṭba'a al-Amīriyya.

Ibn Sīnā. 1970. *Al-Shifāʾ, al-Manṭiq, al-ʿIbāra*, ed. I. Madkūr and M. al-Khuḍayrī. Cairo: Dār al-Kātib al-ʿArabī li'l-Ṭibā'a wa'l-Nashr.

Ibn Sīnā. 1959. *Avicenna's De Anima, Being the Psychological Part of Kitāb al-Shifāʾ*, ed. F. Rahman. London: Oxford University Press.

Ibn Sīnā. 1974. *The Life of Ibn Sina: A Critical Edition and Annotated Translation*, edited and translated by William E. Gohlman. Albany: SUNY Press.

Ibn Sīnā. 2009. *The Physics of The Healing*, edited and translated by Jon McGinnis. Provo, UT: Brigham Young University Press.

Margoliouth, D.S. 1905. The Discussion between Abu Bishr Matta and Abu Saʿid al-Sirafi on the Merits of Logic and Grammar. *Journal of the Royal Asiatic Society of Great Britain and Ireland* 37: 79–129.

Secondary Works

Black, Deborah. 2008. Al-Fārābī on Meno's Paradox. In *In the Age of al-Fārābī: Arabic Philosophy in the Fourth/Tenth Century*, ed. Peter Adamson, 15–34. London: The Warburg Institute.

Daston, Lorraine, and Peter Galison. 2010. *Objectivity*. New York: Zone Books.

Hallaq, Wael B. 1997. *A History of Islamic Legal Theories: An Introduction to Sunnī uṣūl al-fiqh*. Cambridge: Cambridge University Press.

Hanson, Norwood R. 1958. *Patterns of Discovery: An Inquiry into the Conceptual Foundations of Science*. Cambridge: Cambridge University Press.

Janssens, Jules. 2004. 'Experience' (*tajriba*) in Classical Arabic Philosophy (al-Fārābī–Avicenna). *Quaestio* 4: 45–62.

Kaukua, Jari. 2015. *Self-Awareness in Islamic Philosophy: Avicenna and Beyond*. Cambridge: Cambridge University Press.

Knorr Cetina, Karin. 1999. *Epistemic Cultures: How the Sciences Make Knowledge*. Cambridge, MA: Harvard University Press.

McGinnis, Jon. Demonstrating Experience: Avicenna's Solution to Meno's Paradox and Its Implications for the Sciences. *The Monist*, forthcoming.

McGinnis, Jon. 2003. Scientific Methodologies in Medieval Islam. *Journal of the History of Philosophy* 41 (3): 307–327.

McGinnis, Jon, and David Reisman. 2007. *Classical Arabic Philosophy: An Anthology of Sources*. Indianapolis: Hackett.

Percy, Walker. 2000. Bourbon. In *Signposts in a Strange Land: Essays*, ed. Patrick Samway, 102–107. New York: Picador.

Pollock, John L. 1987. Epistemic Norms. *Synthese* 71 (1): 61–95.

Shihadeh, Ayman. 2015. *Doubts on Avicenna: A Study and Edition of Sharaf al-Dīn al-Masʿūdī's Commentary on the Ishārāt*. Leiden: Brill.

Siddiqui, Sohaira Z.M. 2019. *Law and Politics under the Abbasids: An Intellectual Portrait of al-Juwayni*. Cambridge: Cambridge University Press.

Jon McGinnis is a professor at the University of Toronto. He is the author of *Avicenna* (Oxford University Press, 2010), translator and editor of Avicenna's *The Physics* (Brigham Young University Press, 2009), and the author of numerous articles on medieval philosophy and science.

Correction to: Revisiting Premodern Islamic Science and Experience

Hannah C. Erlwein and Katja Krause

Correction to: H. C. Erlwein and K. Krause, *Revisiting Premodern Islamic Science and Experience*, SpringerBriefs in History of Science and Technology, https://doi.org/10.1007/978-3-031-76085-3

In the original version of the book, the following belated corrections have been incorporated in the copyright page:

The Open Access funding information has been updated as follows in the copyright page: Open Access funding is provided by the Max Planck Society and the Max Planck Institute for Comparative and International Private Law. The MPDL Logo has been placed in the front cover and title page.

The book has been updated with the changes.

The updated version of this book can be found at
https://doi.org/10.1007/978-3-031-76085-3

H. C. Erlwein and K. Krause, *Revisiting Premodern Islamic Science and Experience*, SpringerBriefs in History of Science and Technology,
https://doi.org/10.1007/978-3-031-76085-3_6

Index

H. C. Erlwein and K. Krause, *Revisiting Premodern Islamic Science and Experience*, SpringerBriefs in History of Science and Technology,
https://doi.org/10.1007/978-3-031-76085-3